Science
PATHWAYS of DISCOVERY

Science
PATHWAYS of DISCOVERY

Edited by Ivan Amato

John Wiley & Sons, Inc.

This book is printed on acid-free paper. ♾

Published by John Wiley & Sons, Inc., New York
Published simultaneously in Canada

Design and production by Navta Associates, Inc.

ISBN 0-471-05660-X

Printed in the United States of America

10 9 8 7 6 5 4 3 2 1

CONTENTS

Science
PATHWAYS of DISCOVERY

INTRODUCTION

The greatest gifts science has offered humanity over the past half-millennium are new reasons for feeling awe.

To be sure, the fundamental mystery of existence has been an underwriter of awe since the get-go. That there is something instead of nothing is perhaps the most obvious and most awesome fact of life. But the ever more sophisticated scientific understanding of this existence builds structure into this primary and otherwise amorphous source of awe.

Ever since Galileo pointed his telescopes upward, for one, natural philosophers and then astronomers have revealed the night sky as a portal to a universe so vast, magnificent, and strange that it strains the credulity even of us moderns. Black holes and neutron stars ought to be deemed preposterous, the hallucinatory visions of madmen. But they're real. Pity our prescientific progenitors whose mythic explanations of the night sky remained unembellished by such fantastic phenomena.

Pity them also for what they didn't know about biology and the material world. Consider how much more awesome life has become now that several centuries of biology has uncovered the intricate dynamics that connect an entire living kingdom to the exquisite cellular and molecular dance that underlies it all. Consider how much more texture and variety chemists have added to the world as they've learned the ways of the periodic table—the mother of any kind of naturally occurring or manmade stuff that ever was, is, or can be.

The lineage of ***homo sapiens scientificus*** has shown that the story of a universe can be extracted from a grain of sand. It has revealed how the comprehensive diversity of phenomena—ranging from quarks to atoms to molecules to cells to mosquitoes and snow and people and planets and stars and perhaps even multiple universes as incomprehensibly enormous as the one we inhabit—share common lawful bases even as each individual phenomenon remains unique. Hydrogen and oxygen make water; but water is completely different from its ingredients. Minds emerge from cells and molecules; but the ephemeral and psychological qualities of mind-phenomena seem to bear no resemblance to the traits of cells and molecules.

I was reminded twelve times over of science's gift of awe while editing *Science* magazine's *Pathways of Discovery* essay series, which appeared on a one-essay-per-month basis throughout the year 2000. Published by the American Association for the Advancement of Science, or AAAS, *Science* is best known as one of the most visible and influential venues for publishing original research. Besides technical research reports and articles, each of its weekly issues also provides readers with news, essays and commentary, written variously by professional journalists, working scientists, policymakers, and other players in the scientific world. The *Pathways* series—which began as an idea during an editorial meeting in late 1998 presided over by then editor-in-chief Floyd Bloom—was unlike anything that had been in the magazine before.

Written by 16 highly accomplished, decorated, and articulate scientists, the essays take readers through a recapitulation of the historical, personal, and intellectual pathways that led to many of today's most exciting and influential areas of investigation. From quantum physics to atmospheric science, genomics to neuroscience, the essays recount how each field came to prominence, what that field is like today, and how it might evolve in the near future.

The first essay, generously and brilliantly penned on short order (during the 1999 World Series, no less) by paleontologist, essayist, and baseball aficionado Stephen Jay Gould, sets the stage for the topical essays that follow by addressing the question of what characterizes the scientific way of knowing the world. Gould does so, in part, with a surprising historical tale, involving anatomically suggestive and deceptive fossils known as hysteroliths (you will have to read the essay to find out more about these), which illustrates how scientific fact and interpretation evolve within social and cultural contexts that feed back into those very facts and interpretations. In so doing, Gould tries to undermine much of the foundation for today's so-called science wars between "realists," who claim that science moves progressively toward objective truth, and "relativists," who say all truth-claims, including scientific ones, are ineluctably dependent upon the cultural context in which they are made.

In a hundred different ways, the next eleven essays illustrate how realists and relativists are both right. Each one tells a tale rich with concepts and personalities, breakthroughs, setbacks, and, more than anything, discovery. What's more, each essay rewards readers with a chance to be awed by what we all can know about ourselves, our world, and our universe, because scientists have bothered to ask questions and seek answers.

In his essay on planetary science, for example, David Stevenson of the California Institute of Technology treats readers to his take on the nearly 500-year-long adventure that has revealed an exotic and magnificent menagerie of planets and moons in our local cosmic neighborhood. And as planetary scientists bushwhack into one of the newest phases of their discipline—the discovery of planets and planetary systems outside of our own solar system—Stevenson reminds his readers of one of the most awesome thoughts possible: that there likely are other inhabited worlds out there.

In their essay, coauthors Eric Lander and Robert Weinberg usher readers' thoughts inward along a journey to the center of biology that opens up another vast and awesome window. In an almost impossible combination of conciseness and readability, they recapitulate the pathways that led in 2000 to the near completion of the sequence of the human genome. The prospects are, well, awesome. In their words:

> The availability of the complete parts lists of organisms, that is, catalogs of all their genes and thus all their proteins, has been redirecting biologists toward a global perspective on life processes—to study the role of all genes or all proteins at once . . . Biology enters this century in possession, for the first time, of the mysterious instruction book first postulated by Hippocrates and Aristotle.

Neuroscientist Eric Kandel, who learned of his 2000 Nobel Prize one week after he and his coauthor, Larry Squire, contribute what may be one of the most concise and authoritative accounts of how scientists have discovered the many intimate links among molecules and cells, and brains and minds. It is no great leap to the truly awesome realization that the very minds that emerge from this linkage are capable of discovering how it is that they can know from whence they came—and how it is that they can know anything at all.

Each essay in this book becomes an invitation to be awed about the world from a different angle. Science has evolved and diversified in countless ways. So there is no pretense here that a dozen essays can provide more than a few pixels of the portrait of nature that scientists continue to paint with ever more resolution and scope. That there is so much to be awed about in even the small portion of the scientific world covered in this book, however, only illustrates how far scientists have traveled.

Many thanks are in order. Although the words of the scientist-essayists themselves are quite visible in this book, the generosity and passion with which these contributors composed and offered those words is less apparent and so is herewith acknowledged. Also, just as the scientific enterprise has always depended on standards of measurement, means of communication, and other often hidden infrastructural elements, so too did this series of essays depend on foundations invisible to readers. If it weren't for many members of *Science* magazine's art, production and editorial staffs, as well as the support staffs of the authors who responded to numerous requests for pictures, résumés, and countless other items, this series would never have been more than an idea. Similar thanks go to the editorial and production staffs at Wiley who have transformed the original *Pathways* series into this book.

—Ivan Amato

Pathways of Discovery: A Timeline

Along the timeline, numbers and arrows mark points of departure. These are moments in the history of science when the pathways of discovery associated with each of the 11 topical essays might have begun to unfold.

Prior to 600 B.C.

Prescientific Era
Phenomena explained within contexts of magic, religion, and experience.

Sixth century B.C.

Greek philosophers begin constructing natural and rational explanations for phenomena.

582–500 B.C.

Pythagoras
Recognizes connection between mathematics and nature.

"All is number."

Fifth to fourth centuries B.C.

Leucippus and Democritus
Articulate the earliest known version of atomism.

140

Ptolemy
Summarizes his geocentric astronomy, which reigns until Copernicus (1543).

131–200

Galen
Writes text on anatomy that is preeminent until Vesalius (1543).

500

Symbol for "zero" created in India.

1000

Mayan and Hindu skywatchers make astronomical observations for agricultural and religious purposes.

427–347 B.C.

Plato
Emphasizes the value of theory.

384–322 B.C.

Aristotle
Supports logic and common sense in the observation and classification of natural phenomena. His writings become a cornerstone of Western thinking for nearly 2,000 years.

Birth of alchemy.

Third to second centuries B.C.

Euclid
Writes *The Elements,* one of the earliest logically built texts using axioms and logical deduction for geometric proofs.

287–212 B.C.

Archimedes
Recognizes how theory can lead to technological insight.

1166

Oxford University is founded and becomes the model for other European universities.

Thirteenth century

Mechanical clocks proliferate.

1285–1349

William of Ockham
Touts simplicity in explanation as a virtue.

1370

Synchronization of clocks in Paris, an emblematic moment in the history of standardization.

1500s

1454

Johannes Gutenberg
Demonstrates movable type printing press in Germany.

1454

Paracelsus
His anti-authoritarianism helps West break loose from scholastic dogma.

1543

Andreas Vesalius
Publishes *De Humani Corporis Fabrica,* displacing Galen as the authority on anatomy.

1543

Nicolaus Copernicus
Publishes *De Revolutionibus Orbium Coelestium,* displacing geocentric (Ptolemaic) system with heliocentric system.

1609

Johannes Kepler
Publishes *Astronomia Nova,* in which he defines the first two laws of planetary motion.

1610

Galileo Galilei
With one of the first telescopes, he makes extensive astronomical observations; publishes *Sidereus Nuncius.*

2

"I have taken all knowledge to be my province."

1620

Francis Bacon
Publishes *Novum Organum,* in which he espouses careful observation and inductive reasoning as a basis of scientific study.

1629

William Harvey
Publishes *De Motu Cordis et Sanguinis,* by which he establishes the basic geometry of the circulatory system.

1600s

1570

Tycho Brahe
Increases accuracy of astronomical observation.

1590

Zacharias Janssen
Builds first microscope.

1600

William Gilbert
Publishes *De Magnete,* in which he postulates that Earth behaves as a huge magnet.

1608

Hans Lippershey
Manufactures first telescopes; originally used for military purposes.

1637

René Descartes
His *Discours de la Méthode* combines skepticism, algebra, and geometry into the powerful mathematical method of analysis.

"We live submerged at the bottom of an ocean of the element air."

1643

Evangelista Torricelli
Invents the barometer.

10 ▶

1660

Royal Society in England is legally chartered.

1661

Robert Boyle
Publishes *The Skeptical Chymist,* helping to establish a more modern basis for chemistry.

1700s

1676

Anton van Leeuwenhoek
Uses simple microscope to magnify world up to 200 times. In pond water, he sees tiny living organisms. A pathway to microbiology opens.

1686

John Ray
Starts publishing a three-volume classification of 18,600 different plant species.

1687

Isaac Newton
Publishes his *Principia Mathematica,* in which he lays out mathematically the laws of motion, universal gravitation, and the scientific method in general.

1705

Edmund Halley
Predicts that a comet will return in 1758. When it does, it serves as a powerful demonstration of science's ability to understand the world.

1758

Halley's Comet returns.

1759

Kaspar Wolff
Shows that specialized organs develop out of unspecialized tissue, a pathway to developmental biology.

1769

Antoine Lavoisier
Raises practice of making quantitative measurement in chemical studies to a level that earns him eternal credit as the father of modern chemistry.

1780

Luigi Galvani
Observes connection between electricity and muscle action.

1714

Daniel Gabriel Fahrenheit
Invents mercury thermometer and standardizes it with ice and boiling water.

1720

Lady Mary Wortley Montagu
Introduces early technique of immunization.

1735

Carolus Linnaeus
Publishes *Systema Naturae,* thereby greatly pushing forward systematic study and rules of classification of living kingdom.

1750

Benjamin Franklin
Shows lightning to be an electrical discharge.

1790

Metric system developed.
. . . kilometer, meter, millimeter . . .

1795

James Hutton
Publishes *Theory of Earth,* arguing for a picture of gradual, steady terrestrial changes that became known as uniformitarianism.

1796

Edward Jenner
Develops technique he called vaccination; the science of immunology begins

1797

Georges Cuvier
Publishes first work of comparative anatomy that would support the later evolutionists; his future work with fossils would found the field of paleontology.

4 ▶

1800s

1806

Alessandro Volta
Invents the electric battery, thereby opening the way to several new sciences—including electrochemistry—and technologies.

1808

John Dalton
Publishes *New System of Chemical Philosophy,* refining idea of atoms and how they combine.

1809

Jean-Baptiste de Lamarck
Publishes *Zoological Philosophy,* in which he proposes that evolution occurs by the inheritance of traits acquired during lifetime.

1820

Hans Christian Oersted
Discovers connection between electricity and magnetism, that is, the basic principle of electromagnetism.

1840

James Joule
Develops the concept of conservation of energy.

1844

First electric telegraph.

7

1856

William Ferrel
Publishes *Essay on the Winds and Currents of the Ocean,* a pivotal work in history of meteorology.

1859

Charles Darwin
Publishes *The Origin of Species.*

9

1828

Friedrich Wöhler
Synthesizes an organic compound, urea, in the laboratory, thereby breaking down categorical distinction between life and nonlife.

1830

Charles Lyell
Publishes first volume of *The Principles of Geology.*

1837

Louis Agassiz
Argues for the existence of an ice age in the past.

1838–39

Matthias Schleiden and Theodor Schwann
Develop cell theory—that living things are made of cells.

"The elementary parts of all tissues are formed of cells in an analogous, though very diversified manner, so that it may be asserted that there is one universal principle of development for the elementary parts of organisms, however different, and that this principle is the formation of cells."

1860s

Henry Sorby
Emphasizes how material properties depend upon their internal microscopic structure.

1866

Gregor Mendel
Publishes work containing his famous laws of heredity.

1869

Dmitri Mendeleev
Publishes his *Periodic Table of the Chemical Elements.*

1889

Camillo Golgi
Uses techniques to lay bare the cellular anatomy of the brain.

11

1900s

1895

Wilhelm Röntgen
Discovers X rays.

1897

Marie Curie
She and husband Pierre clarify concept of radioactivity as an atomic property.

1897

J. J. Thomson
Discovers the electron.

1900

Max Planck
Quantum science is born.

1929

Edwin Hubble
Shows that the farther a galaxy is from earth the faster it is receding—a cornerstone of the discovery that the universe is expanding.

1930

Linus Pauling
Updates chemistry with ideas of quantum mechanics.

1938

Hans Spemann
Proposes the concept of cloning.

1939–45

Birth of nuclear technology.

6

8 ▶

12 ▶

1905

Albert Einstein
Develops special theory of relativity; later (in 1916) publishes *General Theory of Relativity.*

1912

Alfred Wegener
Proposes the continental drift theory.

1920s

Quantum mechanics emerges.

1927

Georges Lemaître
Introduces big bang theory.

"It has not escaped our notice that the specific pairings we have postulated immediately suggests a possible copying mechanism for the genetic material."

1947

William Shockley, Walter Brattain, and John Bardeen
Invent the transistor.

1953

James Watson and Francis Crick
Publish the structure of DNA.

1957

Soviets launch *Sputnik.*

1960s

Harry Hess, Dan McKenzie, Jason Morgan, J. Tuzo Wilson, and others
Develop theory of tectonic plates.

1961

Murray Gell-Mann Postulates the existence of quarks.

1962

Rachel Carson Publishes *Silent Spring,* linking pesticide use to ecological ills.

1969

The U.S. Department of Defense's Advanced Research Projects Agency creates ARPAnet, the precursor to the Internet and the World Wide Web.

1970

Hamilton Smith and Daniel Nathans Develop first recombinant DNA techniques.

All-Too-Human Science

Pathways of discovery are only smooth and linear when the nuanced and tortuous history of science is brutally edited to fit into a finite space. The items below represent the countless twists, turns, ironies, contradictions, tragedies, and other unkempt historical details that have synthesized into the far more complex and multitextured reality of the scientific adventure.

Fourth century B.C.: Aristotle rejects the atomism of Leucippus and Democritus in favor of the five-element material system—earth, air, fire, water, and ether. Atomism doesn't surface seriously for 2,000 years. Meanwhile, the five-element philosophy spurs alchemical thought and practice. The power of authority often still poses obstacles to new ideas.

Ninth to fifteenth centuries: The flow of science and technology is mostly into Europe from Islam and China. Among the innovations diffusing westward: paper, chemical technology, glassmaking, the compass, shipbuilding, and gunpowder. In the fifteenth century, the centers of science and technology begin shifting to the West.

1543: Copernicus helps begin the scientific revolution with his heliocentric theory of the universe. He retains the perfect circularity of orbits demanded by Platonic idealism and Ptolemaic astronomy. His work embodies both a radicalism that unlocks new knowledge and a conservatism. These together account for much of the progressive nature of scientific knowledge.

1633: The Holy Inquisition threatens Galileo Galilei with torture unless he renounces heliocentrism. He acquiesces but is confined to house arrest for the rest of his life for previously defying Church authority by publicly promoting the theory. The Galileo affair has since represented—in the most dramatic fashion—the tension between scientific and religious authority.

1970s

The Standard Model of Particle Physics emerges.

1971

First commercial microprocessor manufactured.

1981

Acquired immunodeficiency syndrome (AIDS) recognized.

Mid-1980s

Kary Mullis Develops polymerase chain reaction (PCR).

Seventeenth century: Even as Isaac Newton ascends to become an icon of the scientific revolution, he devotes one million words and countless hours of his intellect to the practice of alchemy. The ability of the human mind to simultaneously entertain contrary belief systems is a perennial source of wonder.

c. 1900: Shortly after Röntgen discovered X rays, the French physicist René Blondlot claimed discovery of what he called "N-rays." Other scientists confirmed his observations. Papers were published. Trouble was, N-rays didn't exist. This episode has become a classic case of "pathological science," in which self-deception, wishful thinking, and selective use and denial of data can lead scientists astray.

1926: Alfred Wegener's theory of continental drift submerges due to the lack of support from his peers (to say the least). Forty years later, moving continents were established as scientific fact. The critical attitude always has been central to the scientific process, but it also can lead to the premature rejection of good ideas. This episode also is testimony to the faith among scientists that the truth will out.

World War II: Often described as "the physicists' war," WW II demonstrated tight links between science, technology, and national security. Synthetic rubber, radar, proximity fuses, and the fission bomb all emerged from the most extensive government-supported R&D program in history. After the war, this new R&D infrastructure evolved into its present form.

1953: Without Rosalind Franklin's crystallographic data of DNA, James Watson and Francis Crick wouldn't have determined that molecule's structure when they did. The two of them and Franklin's senior co-worker, Maurice Wilkins, shared a Nobel Prize for the discovery. Franklin never received official credit for her contribution, and her role was written out of early historical accounts. Franklin represents the struggle for professional parity that women have waged, and continue to wage, in the scientific community.

1985

Hole in the ozone discovered over Antarctica.

1990

Human Genome Project begins.

1992

World Wide Web demonstrated at CERN.WWW.net

1995

Bose-Einstein condensate created in laboratory.

The Twentieth Century—The Janus Face of Science

Science has never been a value-free enterprise, and its double-edgedness expressed itself with a sobering regularity this century. Before there was "better living through chemistry," there were clouds of death-dealing chlorine on the Western Front of World War I. The next world war and the Cold War ushered in the awesome power of nuclear fission and fusion and a new kind of apocalyptic angst. Then came Rachel Carson. She alerted the world to global-scale ecological threats that can come with wide-scale environmental releases of synthetic chemicals. Her point was reiterated in the 1970s and 1980s when industrial activity and products were linked to planet-altering threats like ozone holes and global warming. Now, the growing ability to understand and manipulate our own genetic foundations opens up almost utopian medical possibilities. But that gleamy-eyed future only comes with the threat of unnerving compromises ranging from genetic-based job discrimination to the possibility of literally bumbling with the genetic constitutions of all generations that follow this one. Science has entangled itself within the most important values there are, and that brings with it unprecedented responsibilities for scientists of the next millennium.

1997

Ian Wilmut and Keith Campbell Clone a sheep from adult cells.

2001

Results of public and private human genome projects published in *Nature* and *Science.*

Acknowledgments

Guidance for this timeline came from many books and other documents, as well as from the following historians: Robert Friedel, Naomi Oreskes, Steven Dick, Garland Allen, Spencer Weart, Sara Schechner, and James Fleming. The editors deserve credit for any remaining errors of fact or interpretation.

1

Deconstructing the "Science Wars" by Reconstructing an Old Mold

STEPHEN JAY GOULD

Scientific pathways are rife with twists, turns, switchbacks, dead ends, leaps of faith, flashes of brilliance, and plenty of other features that add drama to those walking these routes. In this essay, Stephen Jay Gould, one of the most prolific and well-known scientist-writers of this generation, mines both the thoroughfares and byways of history to undermine the supposed foundations on which the "two cultures"—loosely demarcated as the humanities and science—carry on their so-called science wars. Is a scientific approach the only way to an objective truth about the world, or are scientific "truths" constrained and molded by the psychological, social, and cultural contexts in which they are generated?

STEPHEN JAY GOULD is the Alexander Agassiz Professor of Zoology and Professor of Geology at Harvard. He also is curator for invertebrate paleontology at the university's Museum of Comparative Zoology and serves as the Vincent Astor Visiting Professor of Biology at New York University. His passion to communicate to an audience far beyond his scientific peers has resulted in numerous engaging books and essays that reveal the all-too-human process of scientific discovery.

1. What If They Gave a War and Nobody Came?

For reasons that seem to transcend cultural peculiarities, and may lie deep within the architecture of the human mind, we construct our descriptive taxonomies and tell our explanatory stories as dichotomies, or contrasts between inherently distinct and logically opposite alternatives. Standard epitomes for the history and social impact of science have consistently followed this preferred scheme, although the chosen names and stated aims of the battling armies have changed with the capricious winds of fashion and the evolving norms of scholarship—as in scientific novelty versus permanent wisdom in the founding seventeenth-century debate of "moderns" (the empirical method for gaining new knowledge) versus "ancients" (Greek and Roman perfection);[1] science versus religion in a favorite trope of late nineteenth-century secularism;[2] and the sciences versus the humanities in the icon for the second half of the twentieth century, C. P. Snow's "two cultures".[3]

At the close of this millennium, the favored dichotomy features a supposed battle called "the science wars." The two sides in this hypothetical struggle have been dubbed "realists" (including nearly all working scientists), who uphold the objectivity and progressive nature of scientific knowledge, and "relativists" (nearly all housed in faculties of the humanities and social sciences within our universities), who recognize the culturally embedded status of all claims for universal factuality and who regard science as just one system of belief among many alternatives, all worthy of equal weight because the very concept of "scientific truth" can only represent a social construction invented by scientists (whether consciously or not) as a device to justify their hegemony over the study of nature.

But all these dichotomies must be exposed as deeply and doubly fallacious—wrong as an interpretation of the nature and history of science, and wrong as a primary example of our deeper error in parsing the complexities of human conflicts and natural continua into stark contrasts formulated as struggles between opposing sides. When we reject this constraining mental model, we will immediately understand why a science war can only exist in the minds of critics not engaged in the actual enterprise supposedly under analysis. The exposure of this particular naked emperor can only recall the wisdom embedded in a familiar motto of recent social activism: "What if they gave a war and nobody came?"—a statement that may seem a bit limp in its ironic humor at first, but that actually embodies a deep insight about the nature of categories falsely judged as natural and permanent, while truly originating as contingent and socially engendered. (The defining model of dichotomous pairing also lies within the set of mental categories falsely imputed to nature's intrinsic order.)

The best scholars have always been able to scrutinize their own foibles, and an antidote to dichotomous pairing has also existed since antiquity as the *aurea mediocritas,* or "golden mean," of Horace and Aristotle. The *Oxford English Dictionary* traces a first English use to Spencer's, *The Faerie Queene* (1590), with an explicit contrast to dichotomous pairing:

> the face of golden Meane.
> Her sisters, two Extremities:
> strive her to banish cleane.

Most falsely dichotomized battles include important aspects of virtue at each pole, if only we can break through the emotion of mutual anathema and move toward literal mediation. The golden mean of the science wars could not be more commonsensical, or more enlightening, in upholding, at the same time and without contradiction, both the continuous social construction *and* the growing empirical adequacy of scientific knowledge. Why did we ever construe such consonant notions as antithetical? Science, as done by human beings, could only be envisaged and practiced within a constraining and potentiating set of social, cultural, and historical circumstances—a variegated and changing context that, by the way, makes the history of science so much more interesting, and so much more passionate, than the cardboard Whiggery of conventional marches to truth over social impediments (the model that scientists devised for self-serving motives and that still permeates the obligatory historical paragraphs of most scientific textbooks). On the other side, who would wish to deny the probable truth value of science, if only as roughly indicated by increasing technical efficacy through time—not a silly argument of naïve realism, by the way, but a profound comment, however obvious and conventional, about the only workable concept of factual reality.

II. Francis Bacon and the Instrument of Instauration

The objectivist myth of science as a fully general method, rooted in observation by minds consciously free of constraining social bias and using universal tools of reason to accumulate reliable knowledge leading toward an increasingly synthesized theoretical understanding of causes, affixes a definitive label upon our profession, as represented by the false dichotomy of the science wars. The conventional hero in English versions of this myth, Francis Bacon (1561–1626), upheld the new birthright of the scientific revolution by asserting a central paradox in his generation's "battle of the books" between classical and modern knowledge: *Antiquitas saeculi, juventus mundi,* or "the good old days were the world's youth." We should not, as the resolution of this paradox proclaims, regard ourselves as callow, and the ancient Greeks as hoary with wisdom that we might learn to emulate but could never surpass—the standard argument that motivated the Renaissance (literally, the rebirth), a movement that did not strive for novelty on a modern scientific model but sought to rediscover the supposedly eclipsed perfection of ancient knowledge. In contrast, the Greeks and Romans lived during the *world's* youth, whereas we represent the graybeards, enjoying time's benefit and seeing farther by standing on the shoulders of earlier giants (to cite Newton's famous phrase, borrowed from a common aphorism in his day).[4] Knowledge accumulates through time; we now know more than ever before and will continue to advance through the empirical methods of a new and developing discipline called science.

Bacon's name therefore became an adjective in the objectivist myth, a symbol for accurate observation, uncluttered and unbiased by theoretical preferences rooted in social constraints. In a famous passage of his autobiography, for example, Darwin described his procedure in devising the theory of natural selection:[5]

> My first note-book was opened in July 1837. I worked on true Baconian principles and without any theory collected facts on a wholesale scale.

As many historians have noted, Darwin did not (and could not, for no one can) proceed in such an empty-headed manner, and his later recollection that he did so in his youth can only represent an imposition of unachievable professional ideals upon a forgotten reality. Darwin may have had no inkling of natural selection when he began, but his notebooks represent an extended mental adventure in the constant test and rejection of sequential hypotheses.[6] In private letters, Darwin often expressed his keen intuition that facts must be ordered and selected as tests of ideas, and that objectivity can only be meaningfully defined by fair recording and by a willingness (even an eagerness) to alter and reject favored hypotheses in the light of such records. In an equally famous letter of 1863, Darwin wrote:[7] "How odd it is that anyone should not see that all observation must be for or against some view if it is to be of any service!"

Bacon's dubious, and wholly undeserved, reputation as the apostle of a purely enumerative and accumulative view of factuality for the source of theoretical understanding in science rests upon the tables for inductive inference that he included in the *Novum Organum,* the first substantive section following the introduction to his projected *Great Instauration.* Bacon, who has never been accused of modesty, had vowed as a young man "to take all knowledge for my province." To break the primary impediment of unquestioned obeisance to ancient authority (the permanence and optimality of classical texts), Bacon vowed to write a *Great Instauration* (or New Beginning), based on principles of reasoning that could increase human knowledge by using the empirical procedures then under development and now called "science."

Aristotle's treatises on reasoning had been gathered together by his followers and named the *Organon* (tool or instrument). Bacon therefore named his treatise on methods of empirical reasoning the *Novum Organum,* or "new instrument" for the scientific revolution. The "Baconian method," as Darwin used and understood the term, followed the tabular procedures of the *Novum Organum* for stating and classifying observations, and for drawing inductive inferences therefrom, based on common properties of the tabulations.

Perhaps Bacon's tables do rely too much on listing and classifying by common properties, and too little on the explicit testing of hypotheses. Perhaps, therefore, this feature of his methodology does buttress the objectivist myth that has so falsely separated science from other forms of human creativity. But when we consider the context of Bacon's own time, particularly his need to emphasize the power of factual novelty in refuting a widespread belief in textual authority as the only path to genuine knowledge, we may understand an emphasis that we would now label as exaggerated or undue (largely as a consequence of science's preeminent success).

Nonetheless, a grand irony haunts the *Novum Organum,* for this work, through its tabular devices, established Bacon's reputation as godfather to the primary myth of science as an "automatic" method of pure observation and reason, divorced from all gutsy and sloppy forms of human mentality, and therefore prey to the dichotomous separations advocated in our modern science wars. In fact, the most brilliant sections of the *Novum Organum*—scarcely hidden under a bushel by Bacon, and well known to subsequent historians, philosophers, and sociologists—refute the Baconian myth by defining and analyzing the mental and social impediments that lie too deeply and ineradicably within us to warrant any ideal of pure objectivism in human psychology or scholarship. Bacon referred to these impediments as "idols," and I would argue that their intrusive inevitability fractures all dichotomous models invoked to separate science from other creative human activities. Bacon should therefore become the primary spokesperson for a nondichotomized concept of science as a quintessential human activity, inevitably emerging from the guts of our mental habits and social practices, and inexorably intertwined with foibles of human nature and contingencies of human history, not apart, but embedded—yet still operating to advance our genuine understanding of an external world and therefore to foster our access to "natural truth" under any meaningful definition of such a concept.

The old methods of syllogistic logic, Bacon argues, can only manipulate words and cannot access "things" (that is, objects of the external world) directly:[8] "Syllogism consists of propositions, propositions of words, and words are the tokens and marks of things." Such indirect access to things might suffice if the mind (and its verbal tools) could express external nature without bias; but we cannot operate with such mechanistic objectivity: "If these same notions of the mind (which are, as it were, the soul of words) . . . be rudely and rashly divorced from things, and roving; not perfectly defined and limited, and also many other ways vicious; all falls to ruin." Thus, Bacon concludes, "we reject demonstration or syllogism, for that it proceeds confusedly; and lets Nature escape our hands."

Rather, Bacon continues, we must find a path to natural knowledge—as we develop the procedure now known as modern science—by joining observation of externalities with scrutiny of internal biases, both mental and social. For this new form of understanding "is extracted . . . not only out of the secret closets of the mind, but out of the very entrails of Nature." As for the penchants and limitations of mind, two major deficiencies of sensory experience impede our understanding of nature: "The guilt of Senses is of two sorts, either it destitutes us, or else deceives us."

The first guilt of *destitution* identifies objective limits upon physical ranges of human perception. Many natural objects cannot be observed "either by reason of the subtility of the entire body, or the minuteness of the parts thereof, or the distance of place, or the slowness, and likewise swiftness of motion."

But the second guilt of *deception* designates a more active genre of mental limitation defined by internal biases that we impose upon external nature. "The testimony and information of sense," Bacon states, "is ever from the Analogy of Man, and not from the Analogy of the World; and it is an error of dangerous consequence to assert that sense is the measure of things." Bacon, in a striking metaphor once learned by all English schoolchildren but now largely forgotten, called these active biases "idols"—or "the Idolae, wherewith the mind is preoccupate."

Bacon identified four idols and divided them into two major categories, "attracted" and "innate." The attracted idols denote social and ideological biases imposed from without, for they "have slid into men's minds either by the placits and sects of philosophers, or by depraved laws of demonstrations." Bacon designated these two attracted biases as "idols of the theater," for limitations imposed by old and unfruitful theories that persist as constraining myths ("placits of philosophers"); and, in his most strikingly original conception, "idols of the marketplace," for limitations arising from false modes of reasoning ("depraved laws of demonstrations"), and especially from failures of language to provide words for important ideas and phenomena, for we cannot properly conceptualize what we cannot express. (In a brilliant story entitled "Averroes's Search," the celebrated Argentinian writer Jorge Luis Borges, who strongly admired Bacon, described the frustration of this greatest medieval Islamic commentator on Aristotle, as he struggled without success to understand two words, central to Aristotle's *Poetics,* but having no conceivable expression in Averroes's own language and culture: comedy and tragedy.)

But if these attracted idols enter our minds from without, the innate idols "inhere in the nature of the intellect." Bacon identified two innate idols at opposite scales of human society—"idols of the cave," representing the peculiarities of each individual's temperament and limitations; and "idols of the tribe," denoting foibles inherent in the very (we would now say "evolved") structure of the human mind. Among these tribal idols of human nature itself, we must prominently include both our legendary difficulty in acknowledging, or even conceiving, the concept of probability, and also the motivating theme of this article: our lamentable tendency to taxonomize complex situations as dichotomies of conflicting opposites.

In a key insight, and explicitly invoking his idols to dismember the myth of objectivity, Bacon holds that science must inevitably work within our mental foibles and social constraints by marshaling our self-reflective abilities to understand—because we cannot dispel—the idols that always interact with external reality as we try to grasp the nature of things. We might identify, and largely obviate, the theatrical and marketplace idols imposed from without, but we cannot fully dispel the cave and tribal idols emerging from within. The influence of these innate idols can only be reduced by scrutiny and vigilance: "These two first kinds of Idolaes [attracted idols of the theater and marketplace] can vary hardly; but those latter [innate idols of the cave and tribe], by no means be extirpate. It remains only that they be disclosed; and that same treacherous faculty of the mind be noted and convinced."

In a striking metaphor, Bacon closes his discussion of idols by describing our scientific quest as an interplay of mental foibles and outside facts, not an objective march to truth—a marriage of our mental propensities with nature's realities, a union to be consummated for human betterment: "We presume ° that we have prepared and adorned the bridechamber of the Mind and of the Universe. Now may the vote of the marriage-song be, that from this conjunction, human aids, and a race of inventions may be procreated, as may in some part vanquish and subdue man's miseries and necessities."

III. Olaus Worm and the Archiater of Darmstadt

We need to invoke Bacon's general model of advancing science, inextricably intertwined with and potentiated by our mental foibles and social constraints, if we wish to fracture the false dichotomy of objective realism versus social constructionism that defines and fuels the illusory science wars. But small and concrete cases debunk this spurious conflict even more effectively by proving that both supposedly opposite poles invariably work together, as science builds genuine items of natural knowledge from constantly changing and persistently indivisible mixtures of observation and socially embedded interpretation. I therefore apply this truly Baconian model of advancing science *within* socially constructed explanatory matrices to a particular, well-bounded case—an exemplification of the most important voyage of discovery ever completed within my profession of paleontology: the sixteenth- to eighteenth-century debate on the nature of fossils (spanning the seventeenth-century "invention" of modern science from the early years of Bacon and Descartes to the consummation wrought by Newton's cohort).

I shall describe a complex transition that occurred over two centuries, from Agricola's first geological treatise of 1546 to Linnaeus's taxonomic compendium of 1735, and that indubitably features the genuine discovery and construction of an objective factual truth about the nature of a puzzling natural object. In a cryptic one-liner in his *Historia naturalis,* Pliny had spoken of stones that resembled female genitalia on one side and the corresponding male parts on the other. Georgius Agricola and Konrad Gesner, the mid–sixteenth-century polymaths and founders of modern paleontology, described stones that corresponded to Pliny's words, and these specimens then assumed the standardized name of hysteroliths (womb stones, or vulva stones), because the female likenesses on one side seemed so much more impressive than the vague male analogies only sometimes found on the opposite side.

The nature of hysteroliths posed a major puzzle throughout the sixteenth and seventeenth centuries. No one regarded them as actual petrified parts of human females, so why did they resemble the human vulva so closely in form? Did the similarity merely represent an accidental *lusus naturae* (sport or joke of nature), or did this morphological correspondence between mineral and animal kingdoms express some causal property that might be utilized for human benefit? By the mid–eighteenth century, all scientists had accepted the discovery that hysteroliths form as internal molds of fossil brachiopod shells, with the slitlike "vulva" corresponding to a ridge on the shell's interior, expressed on the mold as an incision. The resemblance that had inspired the original name must therefore be meaningless and accidental—and the old designation disappeared.

The first published illustration of a hysterolith adorns the famous 1655 museum catalog of the Danish naturalist Olaus Worm.[9] He recognized the specimens as molds of some object but chose not to speculate about the organic versus inorganic status of the original model. He also felt comfortable with a claim for human utility, based on the formal resemblance and suggested by an eminent local physician who had discovered the specimens. Worm wrote (my translation from his Latin original):

Vulva stones. These are the first published illustrations of hysteroliths, alias vulva stones. They appeared in a 1655 museum catalog of the Danish naturalist Olaus Worm.

These specimens were sent to me by the most learned Dr. J. D. Horst, the archiater [chief physician] to the most illustrious Landgraf of Darmstadt. . . . Dr. Horst states the following about the strength of these objects; these stones are, without doubt, useful in treating any loosening or constriction of the womb in females. And I think it not silly to believe, especially given the form of these objects that, if worn suspended around the neck, they will give strength to people experiencing problems with virility, either through fear or weakness, thus promoting the interests of Venus in both sexes *(Venerem in utroque sexu promovere).*

At first glance, and as nearly always interpreted, this story of scientific progress seems to fit the old model of empirical objectivity gradually dispelling the prejudicial darkness of antiquated social belief—the realist side of the science wars. How can we view Dr. Horst's opinion that an inert rock might cure human illness by virtue of an accidental resemblance to the afflicted human parts as anything but the silly superstition of a "prescientific" age?

The conventional and hagiographical history of paleontology affirms this view by presenting a heroic tale of scientific light dispelling ideological darkness in three sequential stages. Consider, as a standard source, the early twentieth-century *Transformations of the Animal World* by the leading French paleontologist Charles Depéret.[10] In the first stage, Depéret writes, nearly everyone viewed fossils as inorganic products of the mineral kingdom, formed by fatuous means that could not be regarded as scientific:

The Middle Ages retained the ideas of Aristotle, and almost unanimously adopted the theories of the spontaneous generation of fossils or petrifactions under varying formulas, such as plastic force, petrifying force, action of the stars, freaks of nature, mineral concretions, carved stones, seminal vapours, and many other analogous theories. These ideas continued to reign almost without opposition till the end of the sixteenth century.

In stage two, advancing science fractured these myths and established the organic nature of fossils. But progress remained stymied by religious dogmas that designated all fossils as relics of Noah's flood:

The seventeenth century saw little by little the antiquated theories of plastic force and of carved stones disappear, and the animal or vegetable origin of fossil remains was definitely established. Unfor-

> tunately the progress of palaeontology was to be retarded for a long space of time by the rise and the success of the diluvian theories, which attributed the dispersion of fossils to the universal deluge, and endeavoured to adapt all these facts to the Mosaic records.

Finally, in stage three, the advancing force of exact description dispels this final impediment and establishes the fossil record as a chronology for an immensely long history of life. The gradual and progressive triumph of objective observation over social and ideological constraint has now been completed.

> Yet there were, among these partisans of the Flood, a few men of worth, whose principal merit, outside their too frequent extra-scientific speculations, was that they deeply studied fossils and spread the better knowledge of them by exact representations. This task of the description and illustration of fossil animals was especially the work of the scholars of the eighteenth century which was the age of systematic zoology. From all quarters they set themselves to gather and collect fossils, to study and describe them by the aid of plates often of great beauty of execution, to which modern palaeontologists are still compelled to have recourse.

This simplistic tale may match some cardboard heroic fantasies about the invincibility of scientific reasoning and the inevitability of human progress. But Depéret's tripartite story fails miserably as an accurate history of a genuine scientific advance for at least three reasons:

1. The three putative stages (fossils as inorganic, fossils as relics of Noah's flood, and fossils as products of a long history of life) cannot represent a progressive chronology, because all three opinions vied as alternatives from the very beginning of recorded paleontology. In the first decade of the sixteenth century, well before the initiating publications of Gesner and Agricola, both Leonardo da Vinci and Girolamo Fracastoro explicitly discussed these three interpretations as the full set of conceivable explanations for the greatest particular problem posed by fossils: How can objects looking so much like modern marine shells get into rocks on the tops of high Italian mountains? Moreover, both Leonardo and Fracastoro personally favored the third alternative, universally accepted today, and initially developed by nearly all classical Greek authors who wrote about fossils (and who generally accepted Aristotle's view of Earth's potential eternity, leading to a cyclical notion of changing positions for land and sea, with modern mountaintops representing ancient sea floors). Leonardo penned his ideas into private

notebooks that had no influence upon the later history of science, but Fracastoro's opinions received prominent attention in several standard seventeenth-century sources[11].

2. The inorganic theory, supposedly first in time and most foolish in content, did not represent the emptiness of pure ignorance based on a failure to observe actual fossils carefully, but rather made eminently good sense in light of a different theory about the nature of reality—a defendable notion (even in the peculiar version of Darmstadt's archiater) under a disparate concept of causality that only became obsolete after the rise of modern science in the seventeenth century. Under the Neoplatonic doctrine of "signatures," all basic forms achieved explicit representation (through different means) in each of nature's three realms: animal, vegetable, and mineral. These correspondences not only recorded nature's inherent harmony and order but also embodied meaningful sympathies with potential curative power. Thus, the vulva form of the mineral kingdom might help to restore its diseased or depleted analog in the animal kingdom. Such signatures and sympathies make no sense, and seem risible, under a different (and clearly more adequate) concept of mechanistic causality that triumphed in the scientific revolution of the seventeenth century. But the anachronism of later dismissal cannot brand superseded ideas as foolish in their own time. Many of our most cherished beliefs, including concepts that we regard as factually established and free of social bias, will no doubt seem just as bizarre to our successors as the inorganic theory of hysteroliths strikes us today.
3. The reinterpretation of hysteroliths from inorganic replicas of human genitalia to internal molds of fossil brachiopods did not occur primarily by the weight of accumulating observation. Rather, this radical revision arose as a logically implied consequence of major changes in underlying world views—a crucial transition in human intellectual history provoked by complex factors rooted as deeply in altered social contexts as in improving empirical knowledge. First of all, the key factor that secured the organic nature of hysteroliths—the downfall of the Neoplatonic theory of signatures—owed little or nothing to advancing observation of fossils, but rather marked the imposition upon paleontology of a novel and revolutionary approach to understanding nature (called modern science) that specified an organic interpretation of fossils as the unavoidable consequence of a new view of causality.

Secondly, the "right answer" of brachiopod molds demanded far more than a simple accumulation of accurate observations, for this factual resolution presupposed a complex ideological shift in the basic taxonomy of "organized things in rocks," an achievement that required the dismemberment of several theatrical and marketplace idols. No distinct word for organic remains even existed (a classic Baconian idol of the marketplace) before the early nineteenth century, for "fossil," from the past participle of the Latin *fodere* (to dig up), initially referred to any organized object extracted from the ground, including crystals, concretions, geodes, stalactites, and other inorganic items of definite and "interesting" form. As their distinctive status became clear, organic objects first achieved separate recognition as "extraneous fossils" (that is, as objects introduced into rocks from the animal or vegetable kingdom), as opposed to "intrinsic fossils," or mineralogical phenomena. Finally, as both the coherence and the importance of organic objects rose to prominence, the general term "fossil" gradually shifted and contracted to designate organic remains alone.

Until scientists drew an unambiguous taxonomic distinction between organic objects on one side and all complex, regular, and intriguing inorganic phenomena on the other, the status of fossils (in the modern meaning of organic remains) could not be resolved, and hysteroliths would remain in cognitive limbo. As a striking example, consider two illustrations from the mid–eighteenth century:[12] first, from a French publication of 1755, the last gasp of a promiscuous taxonomy that included organic remains with inorganic objects of similar or analogous form. As long as scientists classified brachiopod molds that looked like female genitalia with stalactites that accidentally mimicked male genitalia, the proper causal distinctions would remain elusive, and hysteroliths could not be resolved as brachiopod molds. By contrast, Linnaeus's figure, printed at virtually the same time in 1735, forges the requisite taxonomic division and resolves two centuries of debate by placing hysteroliths into an exclusive category with other brachiopods that, as remains of external shells or as internal molds of species without interior ridges to impose vulvalike slits, show no accidental resemblance to female genitalia. Common causal genesis had finally replaced common overt appearance as the basic criterion for taxonomic union.

Once the necessary theoretical shifts had occurred, and hysteroliths became undoubted organic remains, the few persisting questions could be resolved by procedures close to the stereotype of canonically objective observation: Are hysteroliths the internal molds of vegetable nuts or animal shells? As molds of animal shells, do they represent impressions of clams or of brachiopods? As molds of brachiopods, what species within the phylum grow interior ridges that engrave vulvalike slits on their internal molds? But these observational resolutions only cleaned up a few remaining details. The major advances that converted hysteroliths from ambiguous objects of potentially mineral origin to the internal molds of brachiopods arose as consequences of deep theoretical and ideological transitions rooted as much in social, political, and philosophical changes in European life and values as in simple accumulation of accurate factual information about the natural world.

In short, and however modestly this small incident of two centuries may rank in the general scheme of things, the resolution of hysteroliths as brachiopod molds marks a genuine and indubitable gain in accurate factual knowledge about a fascinating item of external reality—and no genre of victory in all the annals of human achievement could possibly be deemed more noble or more sweet (although other achievements in the arts and humanities may surely claim equal merit!). Nonetheless, paleontology resolved this small problem as all increments in the genuine "progress of science" must be won—by a complex and socially embedded construction of new modes for asking questions and attaining explanations. Science advances within a changing and contingent nexus of human relations, not outside the social order and despite its impediments.

IV. Parsing Science within All Human Creativity

If, to state four propositions arising from this chapter and expressing its central argument:

1. Science truly does "progress" in the sense of gaining, albeit in a fitful and meandering way through time, more useful knowledge that, without mincing words, must record an improving understanding of an objective external world: and if science must also be the work of eminently fallible human beings, freighted with predispositions based on complex factors of social context, psychological hope, mental and temperamental construction, and historical circumstance—and if:

2. Francis Bacon, despite his stereotypical status as an apostle of objectivism based on the automaticity of observations and their mental manipulation to reach inductive conclusions, recognized and emphasized (in perhaps his most famous image) the "idols" of mind and social organization that inevitably make science a social enterprise constructed within changing ideological contexts—and if:

3. Close analysis of any apparently simple and linear sequence in factual gain, leading to the solution of a definite empirical problem—as in the transition

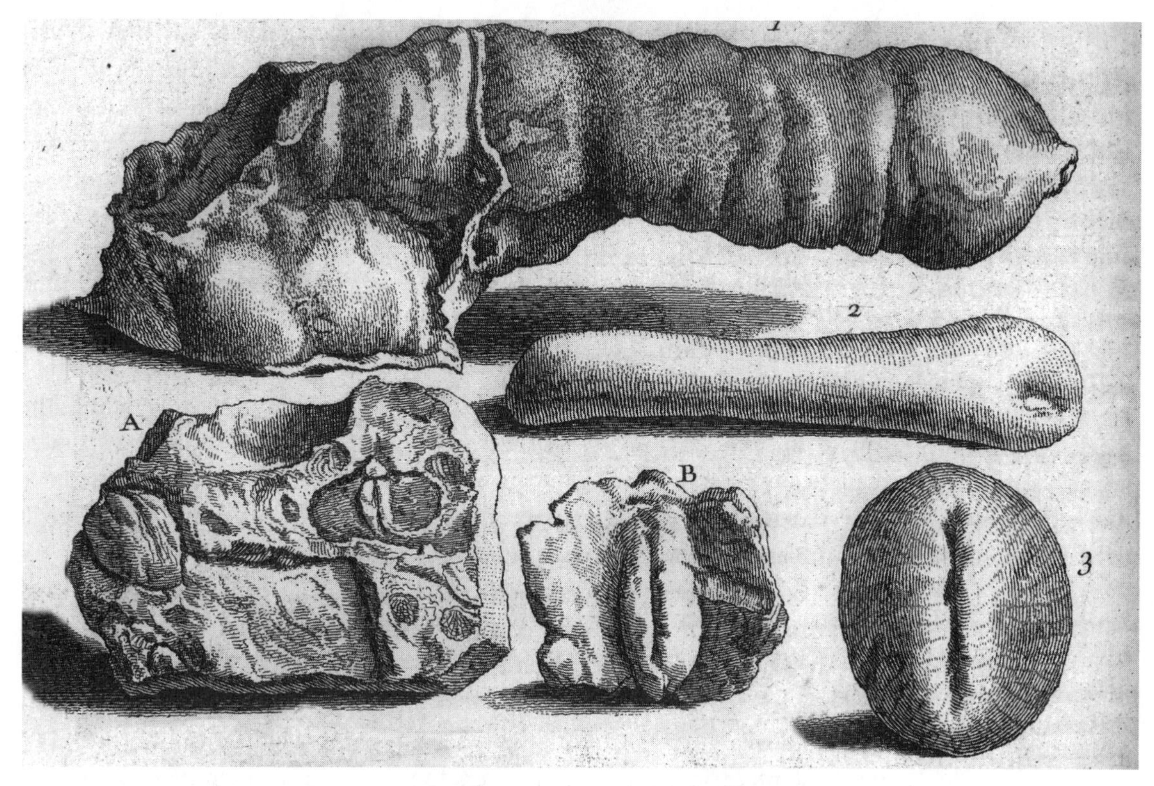

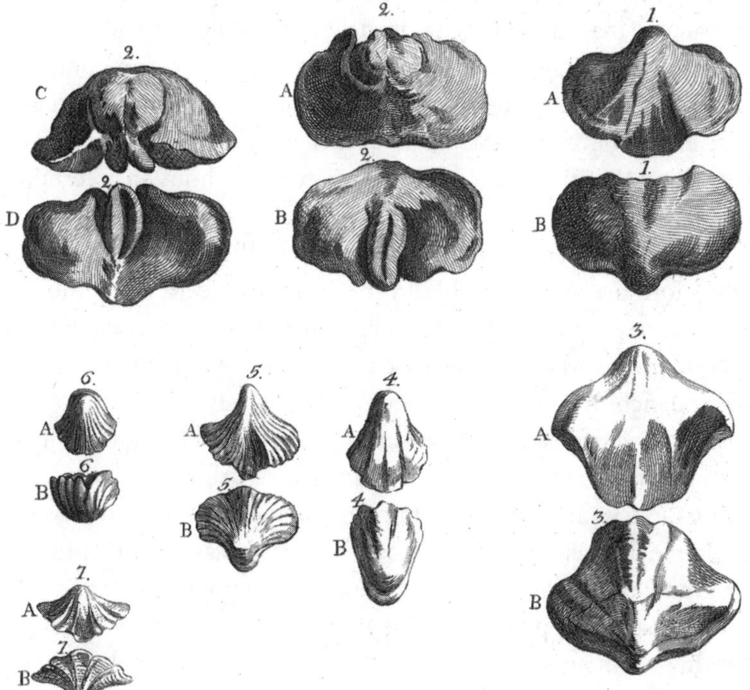

Changing taxonomies. These two sets of illustrations reflect competing mid–eighteenth-century interpretations of hysteroliths. In the plate above (from a 1755 treatise), a variety of natural forms, including stalactites, are classified together because they superficially resemble genitalia. In the plate at left, a contemporary illustration in a work by Linnaeus (1735) forgoes the criterion of common appearances for a basis in common generation, thereby bringing hysteroliths into the narrower class of fossil brachiopods.

from interpreting a prominent group of fossils as "vulva stones" that meaningfully mimic female genitalia to interpreting them as internal molds of certain brachiopods—invariably and ineluctably reveals the central role of social and ideological factors in crucial theoretical shifts that make key observations possible by setting contexts for asking requisite questions—and, finally, if:

4. Dichotomous models of us against them represent Baconian idols of the tribe, or foibles of human mentality imposed upon more complex situations from within—

Then, we must reject the widespread belief that a science war now defines the public and scholarly analysis of this institution, with this supposed struggle depicted as a harsh conflict pitting realists engaged in the practice of science (and seeking an absolute external truth progressively reachable by universal and unbiased methods of observation and reason) against relativists pursuing the social analysis of science (and believing that all claims about external truth can only represent social constructions subject to constant change and unrelated to any movement toward genuine factual knowledge). The very concept of a science war only expresses a basically silly myth, rooted in our propensity for devising dichotomous schemes and supported by the invention of nonexistent, caricatured end-members to serve as straw men in a self-serving rhetorical ploy that can only generate heat without light. (And I do pronounce a plague on this tendency within *both* houses. Social commentators may be more guilty in their frequent mischaracterization of working scientists; but some scientists have constructed equally misleading, and basically philistine, images of social critics out to trash any statement about an ascertainable fact in an objective external world.)

Most working scientists may be naïve about the history of their discipline and therefore overly susceptible to the lure of objectivist mythology. But I have never met a pure scientific realist who views social context as entirely irrelevant, or only as an enemy to be expunged by the twin lights of universal reason and incontrovertible observation. And surely, no working scientist can espouse pure relativism at the other pole of the dichotomy. (The public, I suspect, misunderstands the basic reason for such exceptionless denial. In numerous letters and queries, sympathetic and interested nonprofessionals have told me that scientists cannot be relativists because their commitment to such a grand and glorious goal as the explanation of our vast and mysterious universe must presuppose a genuine reality "out there" to discover. In fact, as all working scientists know in their bones, the incoherence of relativism arises from virtually opposite and much more quotidian motives. Most daily activity in science can only be described as tedious and boring, not to mention expensive and frustrating. Thomas Edison was just about right in his famous formula for invention as 1 percent inspiration mixed with 99 percent perspiration. How could scientists ever muster the energy and stamina to clean cages, run gels, calibrate instruments, and replicate experiments, if they did not believe that such exacting, mindless, and repetitious activities can reveal truthful information about a real world? If all science arises as pure social construction, one might as well reside in an armchair and think great thoughts.)

Similarly, and ignoring some self-promoting and cynical rhetoricians, I have never met a serious social critic or historian of science who espoused anything close to a doctrine of pure relativism. The true, insightful, and fundamental statement that science, as a quintessentially human activity, must reflect a surrounding social context does not imply either that no accessible external reality exists, or that science, as a socially embedded and constructed institution, cannot achieve progressively more adequate understanding of nature's facts and mechanisms.

The social and historical analysis of science poses no threat to the institution's core assumption about the existence of an accessible "real world" that we have actually managed to understand with increasing efficacy, thus validating the claim that science, in some meaningful sense, "progresses." Rather, scientists should cherish good historical analysis for two primary reasons: (1) Real, gutsy, flawed, socially embedded history of science is so immeasurably more interesting and accurate than the usual cardboard pap about marches to truth fueled by universal and disembodied weapons of reason and observation ("the scientific method") against antiquated dogmas and social constraints. (2) This more sophisticated social and historical analysis can aid both the institution of science and the work of scientists—the institution, by revealing science as an accessible form of human creativity, not as an arcane enterprise hostile to ordinary thought and feeling, and open only to a trained priesthood; the individual, by fracturing the objectivist myth that can only generate indifference to self-examination, and by encouraging study and scrutiny of the social contexts that channel our thinking and the attracted and innate biases (Bacon's idols) that frustrate our potential creativity.

Finally, how shall we respond to a harried and narrowly focused scientist who might exclaim:

> Fine, I agree; the history of science may be interesting, relevant, and socially constructed. But I have no time to study such ancillary

> matters, and the results make no practical difference to my life because an objective reality exists "out there," and science would eventually arise to access this factuality in the same basic way, even if our actual history must follow contingent and meandering pathways of social construction.

I would respond that no inevitability attends our eventual understanding of a real world outside our social construction. All basically scientific roads through any conceivable human culture may lead toward an exterior "Rome." But the same Rome shines with different lights tuned to the form and direction of the particular path that people actually construct for their excursion to the eternal city of natural knowledge. We would still know a great deal (perhaps more than now) about the surrounding universe if Zheng He's fifteenth-century ships (five times the length of Columbus's biggest caravel) had continued their explorations beyond Africa, and imperial China had conquered the world. Or if Tamarlane, or Genghis Khan, or Suleiman the Magnificent had vanquished Europe (as each could probably have done, if such issues depended only upon pure technological prowess, and not upon the vicissitudes and contingencies of social practice and personal decision as well). We would still gaze upon Rome, but at what distance, and with what different eyes and concepts?

Not to mention the ultimately sobering thought that just because Rome exists in a position accessible to roads of many forms and styles, no guarantees for human visitation can be located anywhere in the structure of mind or the nature of the universe. We might never have gazed upon this wondrous light in any hue or texture. The dispersal of such false dichotomies as the science wars, and the promotion of science from the heart of its social construction, build a maximally reliable vehicle for this most adventurous of all improbable journeys toward the grandest goal of human striving and natural order.

References and Notes

1. This public and self-conscious struggle culminated in the late 1600s, as older Renaissance convictions about the acme of achievement and textual authority of classical authors ceded to the acknowledgment that newer empirical methods (now called "science") could transcend ancient understanding. In 1704, Jonathan Swift wrote the most famous commentary on this debate, a wickedly satirical essay called, "The Battle of the Books," featuring a war in a deserted library between "ancient" and "modern" volumes.
2. This favored late nineteenth-century dichotomy (still persisting today) viewed social and

technological progress as the outcome of a "warfare" between science and theology, and received a "semi-official" status in two contrasting and phenomenally successful volumes: J. W. Draper's *History of the Conflict Between Science and Religion* (1874) and A. D. White's *A History of the Warfare of Science with Theology in Christendom* (1896). Draper, a vehement anti-Catholic who hoped that liberal Protestant theology might live in peace with science, had presented the speech (on "the intellectual development of Europe considered with reference to the view of Mr. Darwin") that unleashed the famous "debate" between T. H. Huxley and Bishop Samuel Wilberforce in 1860. White, as the first president of Cornell University and a dedicated and ecumenical theist, hoped to persuade his fellow believers that the beneficial and unstoppable advances of science posed no threat to genuine religion but only to outmoded dogmas and superstitions.

3. C. P. Snow, a scientist by training and a novelist and university administrator by later practice, delivered his famous talk on "The Two Cultures and the Scientific Revolution" as the Rede Lecture at Cambridge University in May 1959. He spoke of the growing gap between literary intellectuals and professional scientists, noting for example how "one found Greenwich Village talking precisely the same language as Chelsea, and both having about as much communication with M.I.T. as though the scientists spoke nothing but Tibetan."

4. The great American sociologist of science Robert K. Merton wrote an entire book on pre-Newtonian uses of this image to make the serious point, with a wonderfully light touch, that supposed personal inventions (not claimed by Newton in this case but attributed to him by later commentators) often reflect long and complicated social settings and previous uses. See R. K. Merton, *On the Shoulders of Giants—A Shandean Postscript* (The Free Press, New York, 1965). See also my appreciation of Merton's book: S. J. Gould, "Polished Pebbles, Pretty Shells: An Appreciation of *OTSOG*," in *Robert K. Merton: Consensus and Controversy*, J. Clark, C. Modgil, S. Modgil, Eds. (Falmer Press, New York, 1990), pp. 35–47.

5. F. Darwin, Ed., *The Life and Letters of Charles Darwin, Including an Autobiographical Chapter* (John Murray, London, 1888), vol. 1, p. 83.

6. Darwin's intellectual journey toward the theory of natural selection is brilliantly described, along with a transcription of two key notebooks, in H. E. Gruber and P. H. Barrett, *Darwin on Man* (Dutton, New York, 1974).

7. F. Darwin, Ed., *More Letters of Charles Darwin* (D. Appleton, New York, 1903), vol. 1, p. 195.

8. All the following quotations from Bacon come from his preface to the *Great Instauration*, translated (from the original Latin) by Gilbert Wats: Francis Bacon, *On the Advancement and Proficiencie of Learning: or the Partitions of Science* (Thomas Williams, London, 1674), pp. 17–21.

9. O. Worm, *Museum Wormianum seu Historia rerum rariorum* (Elsevier, Leiden, 1655), p. 84.

10. C. Depéret, *The Transformations of the Animal World* (Kegan Paul, London, 1909), pp. 4–5.

11. At least three seventeenth- and early eighteenth-century museum catalogs describe Fracastoro's views. The earliest reference I can find comes from Andrea Chiocco, *Musaeum Francisci Calceolari Veronensis* (Angelo Tamo, Verona, 1622), p. 407.

12. The 1755 figure of hysteroliths shown with a stalactite resembling male genitalia appears on plate 7 of Dezallier D'Argenville, *L'histoire naturelle éclaircie dans une de ses parties principales, l'Oryctologie* (De Bure, Paris). Linnaeus's accurate picture of 1735 comes from his famous catalog of the collection of Count Tessin: *Museum Tessinianum* (Laurentius Salvius, Stockholm), plate 5.

2

Planetary Science: A Space Odyssey

DAVID J. STEVENSON

Before they were known as Earth's planetary siblings in the solar systems, the most mobile points of light in the night sky were known in Greek as *planetai asters,* or "wandering stars." In this essay, David J. Stevenson recounts the evolution of planetary science. Not only does he chronicle how the field's practitioners have managed to expose the grandeur of our local cosmic neighborhood, but he also reveals how they have become pivotal players in the quest to answer some of the most tantalizing questions possible: How does life arise and where else might it be in the universe?

DAVID J. STEVENSON is the George Van Osdol Professor of Planetary Science at the California Institute of Technology in Pasadena. He received his Ph.D. in theoretical physics at Cornell in 1976, working on the interior of Jupiter, and has since explored all aspects of planetary origin, evolution, and structure, including studies of planet Earth and the moons of the giant planets.

Ancient Times

Many people notice that certain stars move, or wander, more than most of the others, which all seem to move en masse throughout the night. The wandering stars are later recognized as the planets.

1519

Ferdinand Magellan leads first voyage to circumnavigate the globe.

Millennia before anyone realized Earth was only one member of a family of planets orbiting the sun, astute observers noticed a handful of stars that were different. They moved relative to the rest of the stars, which all traveled across the sky as though they were fixed to an enormous rotating bowl. These unusual stars appeared to wander. They became known by the Greek word for wanderers—planets, in English.

Planetary science entered its first great epoch of discovery in the sixteenth century as the scientific revolution itself was developing momentum. Planetary motion was one of the first test-beds for the new physics that was emerging. The second great epoch of discovery, which continues today, began a half-century ago. With the advent of the Space Age in the late 1950s and the continuous development of powerful new tools of observation, planetary scientists have been generating a steady flow of startling revelations, including the recent discovery of planets around other stars, as well as some conditions for life in extraterrestrial settings.

The First Epoch

The first epoch was not lacking in its own startling developments. In the sixteenth and seventeenth centuries, a lineage of celebrity luminaries—most notably, Copernicus, Galileo, Kepler, and Newton—placed the planets (wandering stars) and Earth into the same cosmic category for the first time.

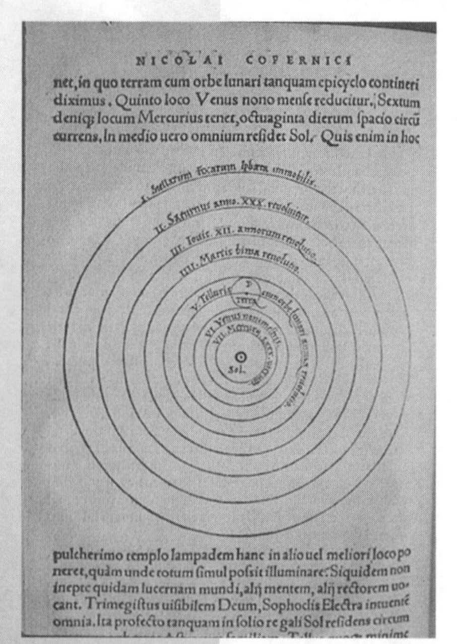

NICOLAI COPERNICI

net, in quo terram cum orbe lunari tanquam epicyclo contineri diximus. Quinto loco Venus nono menſe reducitur. Sextum deniq; locum Mercurius tenet, octuaginta dierum ſpacio circũ currens. In medio uero omnium reſidet Sol. Quis enim in hoc

pulcherimo templo lampadem hanc in alio uel meliori loco poneret, quàm unde totum ſimul poſsit illuminare? Siquidem non inepte quidam lucernam mundi, alij mentem, alij rectorem uocant. Trimegiſtus uiſibilem Deum, Sophoclis Electra intuenté omnia. Ita profecto tanquam in ſolio regali Sol reſidens circum

Revolution. In 1543, Nicolaus Copernicus proposed that Earth and the other planets travel around the sun in his *De revolutionibus orbium coelestium. Jay M. Pasachoff/Visuals Unlimited.*

Galileo—poised between the Copernican bombshell of heliocentrism in the mid–sixteenth century and the culmination of the scientific revolution by Newton more than a century later—was the first modern planetary scientist. In 1609, he quickly grasped the scientific value of a new military technology: the telescope (at first a spyglass for observing naval enemies). With it, he was the first human being to observe another "planetary system"—Jupiter and its moons. The cosmogonic implications of that observation were not lost on him. He also saw mountains on the moon, which instantly entailed the fascinating implication that the celestial bodies could not be objects of perfection and therefore were not the way the ancients proclaimed they had to be.

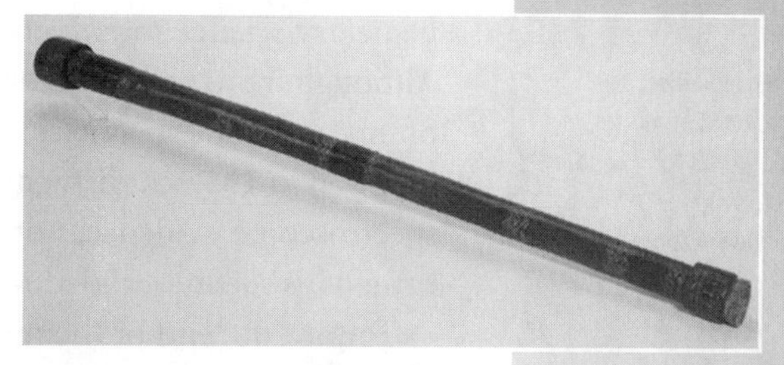

Far-sighted. In 1609, Galileo Galilei built his first telescope. *Istituto e museo di storia della scienza.*

That was a turning point. Even Copernicus's version of the heliocentric system still clung to the Platonic ideal of perfectly circular orbits. Besides Galileo, it would take the combination of Tycho Brahe, who raised the art of astronomical observation to new levels of accuracy, and Johannes Kepler (ironically a Pythagorean at heart)—who tirelessly searched for and found mathematical patterns in Brahe's numbers—to knock those perfect circles of the mind into reality's less pristine, although still orderly, orbits.

During this first epoch of discovery, planets were largely the testing ground for classical physics. Newton's law of gravity and its application to planets demonstrated that physical laws governed not only small, local phenomena such as falling objects, but also large-scale and very distant phenomena as well. Terrestrial and celestial mechanics became facets of the same framework, and the world became more comprehensible.

The first epoch of discovery didn't stop with Newton. Great physicists who came after him continued proving that there was much to learn about physics by observing and explaining planetary phenomena.

In the mid–eighteenth century, Immanuel Kant, and then Pierre-Simon de Laplace[1] again in 1796, postulated the essentially modern view of planets forming from a gaseous disk orbiting the newly forming sun. One of the most celebrated discoveries of the first epoch was the application of celestial mechanics to predict and then observe (in 1846) the presence of a previously unrecognized planet—Neptune.

Celestial mechanics figured prominently because so little was known beyond the massive and obvious presence of the planets. And even those data, for many purposes, had to be thought of as undifferentiated points. To be fair, the telescopes of the time did vaguely reveal atmospheric features of Jupiter and ground features on Mars. But the resolution was too poor for adjudication between competing interpretations of phenomenological descriptions vulnerable to exuberant imaginations.

With data wanting, the power of theoretical approaches came through. The determination of the nature of Saturn's rings stands out as a beautiful illustration of the application of classical physics[2] to solving a problem in

1543
Nicolaus Copernicus proposes that Earth and the other planets travel around the sun in his *De revolutionibus orbium coelestium.*

Late 1500s
Tycho Brahe brings the art of astronomical measurement to new levels of accuracy.

1600
The Inquisition burns Giordano Bruno at the stake, perhaps partly for his belief that Earth revolves around the sun, but also for his belief in other inhabited worlds and his denial of divinity.

1608
Hans Lippershey invents the telescope.

1609
Galileo Galilei builds his first telescope. Johannes Kepler publishes *Astronomia nova* containing the first two laws of planetary motion.

1610
Galileo discovers moons of Jupiter, Saturn's "odd handles," the phases of Venus, and sunspots.

1619
Kepler publishes the third law of planetary motion.

1632
Galileo publishes his *Dialogue* concerning the two chief world systems, but is forced the following year to recant the heliocentric theory under pressure from the Holy Inquisition.

1656
Christiaan Huygens sees and correctly interprets rings of Saturn.

1668
Isaac Newton designs and builds the first reflecting telescope, originally envisioned by James Gregory in 1663.

1671
Jean-Dominique Cassini calculates distances from the sun of all planets then known.

planetary science for which direct observations were then nowhere in sight. Although there had been earlier speculations that the rings may be composed of myriad solid particles, the brilliant physical analysis by James Clerk Maxwell in 1857 confirmed that this was the only possible conclusion. Later, spectroscopic evidence, not to mention radar and direct observations, confirmed his theoretical deduction.

Toward the end of the nineteenth century, the still meager details of Mars that early telescopes could muster relegated discussion of that planet to arguments thin on facts and fat on conjecture. Giovanni Schiaparelli's observations of Mars were meticulous and often documented real features.[3] Yet his use of the Italian word *canali* to characterize networks of long, linear martian features ended up inadvertently fueling an interpretation of these features as artificially constructed structures.

Most prominent in this episode of imagination over reason was the American astronomer Percival Lowell. His late nineteenth and early twentieth century series of popular articles with titles like "Mars as the Abode of Life" reveals as much about the way human brains interpret data and visual information as it does about planets.[4]

The dialectic between theory and observation never goes away. As recently as the 1970s, discussions about the origin of the solar system were still dominated by clever ideas (theory, that is) hardly constrained by the sparse data then available. Observations of disks around nearby stars, primarily by radio telescopes, now allow us to identify the environments in which planets form,[5] and with new optical instruments, including the Hubble Space Telescope and the Keck Telescopes, it is now possible to obtain images of many planets in sufficient detail to allow long-term tracking of atmospheric features and other large-scale changes in planetary appearances. That's the kind of robust data that is required to keep in check the mind's talent for constructing plausible, exciting stories from incomplete bodies of fact.

In the early twentieth century, many astronomers and physicists redirected their attention to things larger and smaller, respectively. Planets receded into the background.

In the pre–Space Age period of the twentieth century—just before planetary science's second epoch of discovery—the two great names in planetary science were Harold Urey and Gerard Kuiper. These scientists reached effectively beyond the severe limitations of visual observation. Urey won the Nobel Prize in 1934 for his discovery of deuterium and called himself a physical chemist, not a planetary scientist. His interest in planets came late, but his legacy is enormous. It is well represented in his book *The Planets*,[6] in which he reveals how the planets' chemistries provide indispensable clues to their formation, structure, and evolution. Urey legitimized planetary science and closely related activities as a serious science complementary to, but distinct from, astronomy.

Many of Urey's conclusions about the nature of the moon and planets are no longer accepted, but his methodology has had a lasting influence. For example, Urey's ideas stimulated his student, Stanley Miller, to carry out experiments on prebiotic synthesis. These have profoundly influenced ideas on the origin of life.[7]

By contrast, Kuiper was part of the astronomy community. He demonstrated the power of spectroscopic observations for determining the composition of planetary atmospheres.[8] Kuiper's approach and conclusions have largely stood the test of time, and the use of spectroscopic techniques on the ground, in Earth orbit, and on spacecraft has proved essential to our understanding of the solar system.

The Second Epoch

The second epoch of planetary science—the period of Great Exploration—began in the latter half of the past century. In just a few decades, scores of spacecraft visited all the known bodies of our solar system at least as large as Earth's moon. And in one of the most audacious feats of humanity to date, a handful of people even walked and buggied around on that once impossibly distant orb.

Driven mainly by the growth of Space Age technology and the Cold War, but also by curiosity, this remarkable second epoch has ushered planetary science far beyond its earlier and mostly supportive role as a test-bed for pioneering physicists. It has evolved into a multidisciplinary endeavor embracing the full richness of planetary processes. Planetary science now has as much in common with geology and the full range of earth science as with astronomy.

The value of getting above all or most of Earth's obscuring atmosphere was recognized early,[9] but the major impetus for the Space Age came from Cold War political considerations. In the United States, the shock of the Soviet Union's successful launch of *Sputnik 1* into orbit in October 1957 instantly mobilized the will and money for an escalating space program. (NASA was founded in 1958.) This led to the gargantuan Apollo missions as well as to much less expensive yet scientifically comparable efforts with unmanned spacecraft.

1687
Newton publishes *Philosophiæ naturalis principia mathematica*, establishing three laws of motion and law of universal gravitation.

1705
Edmond Halley predicts comet (to be known by his name) will return in 1758. It does.

1755
Immanuel Kant proposes solar system arose from vast nebula of material.

1781
William Herschel discovers Uranus.

1795
James Hutton publishes his *Theory of the Earth*, in which he argues for uniformitarianism: All apparent geological features emerge from observable changes unfolding over great expanses of time. Opposing theory, catastrophism, allows for more rapid changes and thereby reconciles better with biblical creation story.

1796
Pierre-Simon de Laplace proposes a nebular hypothesis for the creation of the solar system but goes into more mechanistic detail than Kant did in 1755.

1797
James Hall shows that igneous rock forms crystalline rock upon cooling.

1798
Henry Cavendish determines the mass of Earth: 6.6×10^{21} tons.

1802
William Wollaston discovers dark lines in solar spectrum; twelve years later Joseph von Fraunhofer realizes such lines can be used for spectroscopic studies.

1830
Charles Lyell publishes first volume of his uniformitarianistic tome, *The Principles of Geology*.

1837
Louis Agassiz proposes the idea of an ice age—that at one time glaciers covered Europe.

From the beginning of the rocket and satellite era, scientific discoveries emerged. One of the most important was James Van Allen's discovery of Earth's magnetospheric environment (the Van Allen belts).[10] The influence of these belts (on the operation of electrical power, communications, and other systems) is likely to be felt again as the sun's activity reaches another of its periodic maxima.

The Apollo program was a great technological triumph, but it was also a major scientific event. Interpreting and dating the returned moon rocks brought entirely new kinds of scientists—people previously concerned mainly with meteoritics, earth science, or instrumentation—into planetary research. Even so, the challenge of understanding a planetary body—our own moon, no less—turned out to be a humbling experience even with the wealth of new data. We still do not understand the moon well. And what we have learned about it certainly has not become (as some had hoped) a Rosetta stone for translating the early history of the solar system.[11]

Research conducted during the Apollo program did reveal some aspects of the moon's geological chronology, especially the role of large, frequent impacts in the first half-billion years of the solar system. The lunar rocks also taught us that the moon was severely heated in its earliest times, 4.5 billion years ago. These and other factors underlie much of the currently favored view that the moon arose from a highly traumatic event, perhaps the collision of proto-Earth with a projectile of comparable or greater mass than Mars.[11, 12]

The availability of these artifacts has rendered the moon the only major body in the cosmos (excluding individual meteorites) for which we have a dated chronology of early events. Earth, after all, has obliterated much of the evidence of its early history through the recycling action of plate tectonics. The only other objects for direct study and measurement—rocks and meteorites from Mars or other regions of the solar system—came into our possession via serendipitous, uncontrolled events. Interpreting their meaning has always been plagued by uncertainty about their provenance.

Satellite technology has gone a long way toward countering the dearth of data. Both the United States and the Soviet Union launched early unmanned missions, the crowning achievement of which was the Voyager program. This is not to belittle space-based study of Earth, nor the unmanned spacecraft missions that preceded and followed the Voyager mission, including many missions to other rocky planets (especially Venus and Mars). The Voyager program epitomizes the second epoch of planetary exploration because of the massive amount of high-quality, diverse, and distinct new information it generated.

The mission was a great engineering accomplishment and a remarkable telecommunications feat (a tribute to NASA's Deep Space Network of radio antennas). It was also scientifically spectacular because the mission's two spacecraft visited four planets. *Voyager 1* visited Jupiter (1979) and Saturn (1980); *Voyager 2* visited Jupiter (1979), Saturn (1981), Uranus (1986), and Neptune (1989). The spacecraft also observed some of the planets' moons, which emerged as magnificent bodies with individuality and character as rich and remarkable as the planets themselves.

New images of, and data on, the satellites of Jupiter—which Galileo discovered in the seventeenth century in the first great planetary science payoff of the then-new telescope technology—proved these moons to be a spectacular set of objects:[13] Io is the most volcanically active body in the solar system. Europa has a geologically young and extensively cracked ice surface; we now suspect it has a water ocean beneath its ice shell. Ganymede displays a mixture of terrains, partly tectonic (deformed by internal processes) and partly a record of ancient impacts. And Callisto, which has emerged as a remarkably complex moon in light of data from the 1998 flyby of the still-active *Galileo* spacecraft, ironically reveals no superficial evidence of internal processes.

Saturn and Neptune have their own cast of orbiting characters. Saturn's largest moon, Titan, has a dense atmosphere and a kilometer or so of liquid hydrocarbons.[14] Triton, the large moon of Neptune, has molecular nitrogen frost and plumes of nitrogen mixed with dark (probably carbon-rich) material.[15]

Fraternal twins. Similar in size and bulk, these two moons of Jupiter—Callisto (left) and Ganymede—developed differently. *JPL/NASA/Caltech.*

1846
Johann Galle discovers the planet Neptune based on its predicted position calculated earlier by others.

1857
James Maxwell shows theoretically that Saturn's rings are almost certainly made of myriad small particles that do not coalesce.

1859
Gustav Kirchhoff and Robert Bunsen introduce spectroscopy to chemistry and use it to infer the chemistry of the sun.

1865
Jules Verne publishes the novel *From the Earth to the Moon.*

1877
Giovanni Schiaparelli observes what he calls canali on Mars.

1895
Percival Lowell publishes his book, *Mars,* and argues that the planet is peopled with intelligent creatures that constructed canals and planted crops.

1907
Bertram Boltwood combines information on the half-life of uranium and the proportion of lead found within uranium deposits to estimate age of Earth at 2.2 billion years.

1912
Alfred Wegener proposes idea of continental drift to a chorus of naysayers.

1919
Joseph Larmor develops idea of self-exciting dynamos inside Earth and sun to account for their magnetic fields.

1930
Clyde Tombaugh discovers Pluto.

1931
Harold Urey reasons that hydrogen probably has isotopes and then discovers deuterium spectroscopically, a technique that becomes important for cosmochemical studies.

1937
Grote Reber constructs first radio telescope (9.4 meters in diameter).

The *Voyager* spacecraft also revealed planetary ring systems[16] as distinctive consequences of gravitational physics mixed with fluid dynamics. The rings' seas of particles display wave action as they stream by small moons. And pairs of such moons have a knack for shepherding particles into discrete concentric disks.

The Voyager program also revealed a lot about the dynamics of the giant planets—Jupiter, Saturn, Uranus, and Neptune. All of these have strong, large-scale zonal (East-West) winds which are thought to be fed by smaller-scale motions, but whose ultimate origin and planetary depth remain mysterious.[17] This is a fluid-dynamical challenge that neither resembles conventional meteorology (which addresses the thin atmospheric shell of Earth), nor the much higher-energy, less rotation-dominated convective shell of the sun.

The giant planets also have large magnetic fields, presumably maintained by a dynamo process originating inside the planets where electrically conducting fluids are present. In Jupiter and Saturn, these fluids are thought to arise from at least partial metallization of hydrogen. In Uranus and Neptune, it more likely arises from the mobility of protons in water-rich fluid that is both highly compressed and heated.[18]

More Minds Are Better Than One

Unlike the first epoch of planetary science, the field's second epoch has been less dominated by particular individuals. The subject matter has grown too rich for that. Still, two individuals stand out. Eugene Shoemaker, who died tragically in a car accident in 1997, is one of them. He was a pioneer in understanding the role of impact cratering on solid bodies, helped develop the (still evolving) estimates of surface ages based on these cratering rates, and discovered in 1993—with his wife Caroline and fellow astronomer David Levy—comet Shoemaker-Levy. The following year, the comet collided with Jupiter in one of the most spectacular astronomical events of the past century.

The other standout is the late Carl Sagan. He was a very good scientist, but more important, he was a great communicator. In endeavors such as astronomy and planetary science, which depend heavily on public good will and financial support, people who can convey the value and excitement of the effort are crucial. Sagan also was a champion of the scientific method in an ironic era: Technology and science have great influence on people's lives, yet the general level of understanding of science remains weak enough for magical, superstitious, and other nonrational explanatory frameworks to thrive.

—D. J. S.

I can think of five lessons that have emerged from the second epoch of planetary science.

Lesson One: Common processes are at work.

When we visit other planets with spacecraft for the first time, we're almost always initially surprised by what we find. Yet the underlying physical and chemical processes of these exotic places are not bizarre; their terrestrial analogs are right under our noses. Ice caps form, winds blow, volcanoes erupt, and magnetic fields are produced both here and elsewhere in the solar system. Our far-flung explorations to other planets and moons test our imagination and challenge our basic scientific understanding, but they ultimately confirm our grasp of the basic physics and chemistry as well as expand what we know about the physical universe.

Take Mars. Its mass is one-tenth that of Earth, but it has volcanic structures similar to oceanic island volcanoes on Earth (for example, Hawaii), and it probably has a crustal composition similar to that of Earth's own rock types. Mars also has sand dunes, valley structures similar to those in Earth's arid polar regions, and water-ice polar caps (as well as smaller, mostly seasonal, and decidedly unearthly dry-ice polar caps).

Data returning from the highly successful Mars Global Surveyor mission are showing that the planet has regions of magnetized crust,[19] a testament to an early epoch of martian history when it had an Earth-like magnetic field. In the planet's southern hemisphere, these magnetic regions are organized as East-West stripes. They're somewhat like the magnetic lineations that cover Earth's ocean floor and that helped researchers document sea-floor spreading and plate tectonics. Perhaps Mars also once had plate tectonics and a reversing magnetic field.

1950
Jan Oort suggests that a distant shell of comets surrounds the solar system.

1951
Dirk Brouwer is first astronomer to calculate planetary orbits using a computer.

1957
Soviets launch *Sputnik 1*, the first artificial satellite. Americans follow suit months later in 1958. In 1959, Soviet Lunik satellites reach the moon; one of them crash-lands there.

1958
James Van Allen proves value of satellite-based studies when he uses data from a particle counter on *Explorer IV* to discover Earth's magnetosphere. NASA is created.

1960s
Soviet Union and United States begin what some later call the "Great Epoch" of planetary exploration using satellites, which eventually reach every object in the solar system bigger than the moon.

1969
Human beings land on the moon.

1973
First images of Jupiter transmitted from close vicinity by *Pioneer 10.*

1974
First images of Mercury transmitted by *Mariner 10.*

1976
Viking space probes land on Mars.

1979
Voyager 1 and *2,* the probes of the most successful planetary flyby mission ever, go by Jupiter and some of its moons. The mission ultimately also relays data from Saturn, Uranus, Neptune, and some of their moons.

1986
Space Shuttle *Challenger* explodes soon after launch, killing all seven crew members.

1990
Hubble Space Telescope is placed in orbit. After an initial flaw in the optics is repaired, the HST produces a nearly relentless stream of spectacular imagery.

Currently, these hypotheses are controversial. But they illustrate an important theme. Terrestrial experience and observations provide ground truth about what is possible; extraterrestrial experience and observations test and challenge our ability to extend this base to other planets where geophysical circumstances and history are different.

Lesson Two: Common processes yield diverse outcomes.

Stars are simpler than planets. That enables astronomers to develop powerful organizational principles like the Hertzsprung-Russell diagram with which they can map the intrinsic brightness (luminosity) and spectral class (temperature) of the plethora of stars and then use the resulting graph to identify the main sequence and standard paths of stellar evolution as a function of mass.

Nothing similar exists for planets, because mass, compositional class (rock, ice, or gas), and distance from the sun are not sufficient for characterizing planetary behavior. There are too many degrees of freedom, some of which seem minor yet prove to be major. That's why planetary scientists have come to appreciate that common processes often do not lead to similar outcomes. The richness of outcomes that can develop from the same underlying processes—famously illustrated by millions of biological species descended from the same processes of evolution—is the basis for much of the surprise scientists experience when they first encounter new planets up close.

Consider the role of water on Earth. We do not know where our planet's water came from,[20] the total amount of that water, nor how much of that initial total the planet still has. We have no reason to believe this amount is deterministic, that is, that a planet like Earth equivalently located with respect to its local star will have the same amount of water as Earth does. We do know Earth's water profoundly affects global dynamics. For one, it softens mantle rocks, thereby preparing the way for an asthenosphere—the soft layer underlying the plates.[21] It also quite likely is a pivotal condition for plate tectonics, which in turn partly determines the cycle of water: When plates subduct, water is carried into Earth's interior.

Had Earth started out differently, say, with a modest difference in its amount of water, the planet might have evolved quite differently. And perhaps the major reason Venus is unlike Earth is because it lacks water in its upper mantle. That could at least partly account for the absence of plate tectonics there.[22]

Consider also the role of sulfur, another minor constituent of the Earth-like planets (Mercury, Venus, Earth, and Mars). Sulfur is iron loving, so it likes to be in the iron core of a planet. It's also an antifreeze, so a planetary core with lots of sulfur is less likely to fully solidify. A core that only partly solidifies yields a buoyant fluid and keeps energy available for sustaining the internal material motions required for generating a magnetic field. Two otherwise identical planets with only modest differences in the sulfur concentrations of their cores, therefore, could differ dramatically: One might have a magnetic field and the other might not.[23]

Ganymede and Callisto, two of Jupiter's moons, have ended up remarkably different, even though they are similar in size and bulk composition.[12] Ganymede has tectonic features on its surface; Callisto doesn't. Ganymede's structure is fully differentiated; Callisto's is probably not. Ganymede probably has an Earth-like, dynamo-generated magnetic field; not so for Callisto. The bases of these differences are not understood.

Finally, consider the recent suggestion that giant planets in other planetary systems may have experienced large orbital migrations. That has led to speculation that similar events could have transpired in our own solar system.[24]

We have been at this business long enough that we should appreciate the central role that chance plays in planetary matters. Our solar system has played out one of many possible scenarios in a vast historical progression that has largely deterministic rules, yet whose outcomes cannot be precisely determined. We may now know most of the rules governing planetary phenomena, but we have only begun to figure out and observe the many possible outcomes.

1992
Alexander Wolszczan and Dale Frail discover two Earth-sized planets orbiting a pulsar.

1994
Comet Shoemaker-Levy crashes into Jupiter.

1995
Michel Mayor and Didier Queloz discover the first planet around a sunlike star—51 Pegasi. An era of extrasolar planetary discovery begins in earnest.

1999
Two Mars probes fail, throwing NASA's strategy of faster, cheaper, better missions into question; evidence for more extrasolar planets accumulates.

Lesson Three: The cosmic environment matters.

The influence of the sun and moon on the Earth is as clear as daylight and tides, but we have also discerned less obvious external influences in the history and evolution of a planet. A massive impact at the end of the Cretaceous period on Earth—once controversial but now widely accepted—is a likely cause of the extinction of many species, dinosaurs among them. Very probably, impacts were important in establishing the early environments on Earth and Mars and thereby in abetting or hindering the conditions necessary for life.[25] Jupiter too may have been an unwitting nurturer of life on Earth by restricting the number of impacting bodies reaching here.[26]

The cosmic environment exerts other, less dramatic influences. The gravitational pull of other planets leads to small disturbances of Earth's orbit and orientation, for example. Those, in turn, help determine fluctuations in Earth's climate, including the coming and going of ice ages. The same probably holds true for Mars.[27]

Our cosmic environment is also manifest in the materials around us. It has been established that within some meteorites there are small particles which predate the Earth and our solar system—they came from interstellar space and survived the collapse of the interstellar cloud from which the Sun and planet formed.

Lesson Four: A historical perspective is essential.

A major goal of planetary science is understanding how things came to be. Astronomers may look at large redshifts for clues to ancient events that occurred far away, but planetary scientists trying to understand the solar system look to clues in the rocks and morphological forms on solid planets. This is the geological approach. It embraces the idea that we can read history in the solid bodies around us because they (sometimes) retain a "memory."

Consider the precise dating of meteorites. That approach has consistently revealed an early, relatively brief period of activity (including melting) in the solar system's small, solid bodies that have come our way. The uniformity of these dates is one line of evidence enabling us to speak of the "age" of the solar system (now believed to be about 4.6 billion years).[11] Evidence for extinct radionuclides (in the form of their daughter elements) in meteorites provides an independent check, because the half-lives of those radionuclides are well known. The oldest rocks on the moon are almost as old as the solar system itself, which attests to the rapidity of some planetary processes early on.

Planetary scientists almost always work with a less complete set of data than they would like for reconstructing events. Not even the Apollo astronauts could function as field geologists would. And the technical challenges of getting to and studying Mars mean it may be a long time before a true stratigraphy can be constructed for the various regions of that planet.

Still, there are ways to tease out history. The more limited techniques of photogeology coupled with geophysical modeling have enabled us to estimate the ages of planetary surfaces and the sequences of events on those surfaces. On Venus, for example, the paucity of impact craters leads to the inference of a recycling of the outermost layers in the past billion years. Of course, the apparent lack of any current process capable of this recycling[22] will need to be reconciled with this inference. Perhaps Venus had plate tectonics but no longer does. On Mars, the younger-appearing northern terrains may indicate an early recycling, also suggesting a (more ancient) epoch marked by plate tectonics.[28] On Triton, Neptune's large moon, the crater density indicates that resurfacing is at work. That suggests that Triton may have an active interior, which would be remarkable for a body so small.[29]

Lesson Five: Ground-based data are essential.

It is easy to be impressed by the data returned from spacecraft. It is easier to forget that much of what we learn about planets comes from ground-based activities—either as independent efforts or as complements to spacecraft-based activities. Some of this is from large telescopes, but much of it is small science (bench-top experiments on the properties of materials and computer simulations of the formation of planets and moons).

The science return per dollar invested in ground-based work is very high. Examples include the existence of Jupiter's large magnetic field; the rotation states of Venus and Mercury; the strong greenhouse effect on Venus; the diversity of shapes, rotation states, and compositions of asteroids; the strange surface of Titan; and the persistence and high temperatures of volcanism on Io. What's more, confidently interpreting spectra to learn about the compositions of other atmospheres or interpreting planetary compositions and behavior from condensed matter physics is only possible with laboratory data for comparison.

We have learned that an interdisciplinary approach works best. This is both a weakness and a strength. Planetary science is not a scientific discipline in the usual sense; it is instead a combination of all areas of science that may help practitioners of the field understand how planets work. Outsiders sometimes perceive the resulting science as lacking in the detail and precision apparent in the contributing disciplines. Planetary scientists cannot function like field geologists on other planets (yet!). Consequently, they must rely heavily on physical reasoning, inferential arguments, and modeling for interpreting landforms whose natures are particularly difficult to discern without ground truth. Planetary scientists thus are big fans of computational studies, which have emerged over the past few decades as a widely used, third branch of scientific investigation in addition to the traditional pair of experiment and theory.

There is a great bonus that comes with this broader approach: Planetary scientists have become indispensable players in the quest to answer fascinating questions that would fall outside a more narrowly focused discipline. For one, research into many aspects of Earth's evolution and behavior requires the planetary perspective. And one of the grandest scientific mysteries of all—the origin of life—is unlikely to be solved only by biologists, physicists, and chemists. That accomplishment is sure to require the mindset of planetary scientists as well.

The Future of Planetary Science

The field's future development is likely to emerge from three intertwined trends:

1. the search for extrasolar planets;
2. the search for life elsewhere and for life's origins; and, perhaps most importantly,
3. the search for a fully integrated view of planets in general and our planet in particular.

The first trend brings planetary science back to its roots in mainstream astronomy. The second links planetary science to biology in tackling what arguably might remain the most fundamental unsolved problem of all science—the origin of life. The third trend identifies what is special about planetary science: Conventional disciplines have proven ill-suited for understanding the enlarging diversity of planetary phenomena, and the field has been fostering a holistic approach. (This is all exciting stuff, but it's also expensive and fraught with the perils of overpromotion and underperformance.)

The next portion of planetary science's ongoing second epoch will differ significantly from the one carried out from 1960 to 2000. The heady, first-time excitement of the Great Exploration can never be repeated, but many exciting missions are scheduled or planned by scientists around the world. The Russian space program has waned, but the Japanese and European efforts in planetary science are growing.

The startling, exciting, and increasingly successful quest to find planets around other stars[30] is sure to stretch our minds in the coming years. As this emerging subfield of extrasolar planetary science proceeds beyond the study of mere gravitational influences and points of light to spectra and even onward perhaps to images of those other worlds, planetary science will be revolutionized anew. Much as the telescope enabled Galileo and his contemporaries to forever change the way they and their descendants perceived the night sky, so too might vivid views of extrasolar planets fully rekindle a general feeling for the awesome richness of possibilities in the universe. And we can only conjecture how we all will change if our investigations lead us to but one more humble example of the development of life.

References and Notes

1. This and other aspects of the history are well covered in S. G. Brush, *A History of Modern Planetary Physics* in three volumes (Cambridge University Press, New York, 1996). See also R. E. Doel, *Solar System Astronomy in America: Communities, Patronage and Interdisciplinary Science, 1920–1960* (Cambridge University Press, New York, 1996); R. Schorn, *Planetary Astronomy from Ancient Times to the Third Millennium* (Texas A&M University Press, College Station, 1998).
2. Described in A. F. O'D. Alexander, *The Planet Saturn* (Dover, Toronto, 1962), p. 186.
3. Early observations of Mars are summarized by H. H. Kieffer, B. M. Jakosky, C. W. Snyder in *Mars,* H. H. Kieffer et al., Eds. (University of Arizona Press, Tucson, 1992), pp. 1–33.
4. W. G. Hoyt, *Lowell and Mars* (University of Arizona Press, Tucson, 1976); W. Sheehan, *Planets and Perception: Telescopic Views and Interpretations, 1609–1909* (University of Arizona Press, Tucson, 1988).
5. A. I. Sargent and W. J. Welch, *Annual Review of Astronomy and Astrophysics* 31, 297 (1993); see also V. Mannings, A. Boss, S. Russell, Eds., *Protostars and Planets IV* (University of Arizona Press, Tucson, in press).
6. H. C. Urey, *The Planets: Their Origin and Development* (Yale University Press, New Haven, Connecticut, 1952).
7. S. L. Miller, *Science* 117, 528 (1953); S. L. Miller and H. C. Urey, *Science* 130, 245 (1959).
8. G. Kuiper, Ed., *The Atmospheres of the Earth and Planets* (University of Chicago Press, Chicago, 1952).
9. J. L. Greenstein, in *The Atmospheres of the Earth and Planets,* G. Kuiper, Ed. (University of Chicago Press, Chicago, 1952), p. 112.
10. This essay cannot possibly do justice to the development of "space physics" (the term commonly used to describe magnetospheric and space plasma science). For information on this area, including a historical perspective, see for example, M. G. Kivelson and C. T. Russell, Eds., *Introduction to Space Physics* (Cambridge University Press, Cambridge and New York, 1995).
11. For this and many other aspects discussed here, a fuller discussion and bibliography can be found in S. R. Taylor, *Solar System Evolution* (Cambridge University Press, Cambridge and New York, 1992).
12. The latest work in this area is summarized in R. Canup and K. Righter, *Origin of the Earth and Moon* (University of Arizona Press, Tucson, in press).
13. Recently reviewed by A. P. Showman and R. Malhotra, *Science* 286, 77 (1999).
14. See *Planetary and Space Science* 46 (September and October 1998) for a special issue summarizing current knowledge of Titan.
15. A. P. Ingersoll, *Nature* 344, 315 (1990).
16. P. Goldreich and S. Tremaine, *Ann. Rev. Astron. Astrophys.* 20, 249 (1982); D. N. C. Lin and J. C. B. Papaloizou, *Ann. Rev. Astron. Astrophys.* 34, 703 (1996).
17. T. E. Dowling, *Ann. Rev. Fluid Mech.* 27, 293 (1995).
18. T. Guillot, *Science* 286, 72 (1999).
19. M. H. Acuña et al., *Science* 284, 790 (1999); J. E. P. Connerney et al., *Science* 284, 794 (1999).
20. It is unlikely that water could condense (even as hydrated minerals) at Earth orbit or even Mars orbit at the time of planet formation, so water must have been delivered from greater

distances, perhaps as ice-bearing planetesimals from the zone of Jupiter formation. Comets are no longer considered the dominant source of Earth's water, because they have a D/H ratio that is twice that of Earth's oceans. See for example D. Laufer, G. Notesco, A. Bar-Nun, *Icarus* 140, 446 (1999).

21. G. Hirth and D. L. Kohlstedt, *Earth Planet. Sci. Lett.* 144, 93 (1996).

22. Recent reviews include G. Schubert, V. S. Solomatov, P. J. Tackley, D. L. Turcotte, in *Venus II,* S. W. Bougher, D. M. Hunten, R. J. Phillips, Eds. (University of Arizona Press, Tucson, 1997), p. 1245; and F. Nimmo and D. McKenzie, *Ann. Rev. Earth Planet. Sci.* 26, 23 (1998). For a somewhat different perspective, see S. C. Solomon, M. A. Bullock, D. H. Grinspoon, *Science* 286, 87 (1999).

23. D. J. Stevenson, *Rep. Prog. Phys.* 46, 555 (1983).

24. A low-temperature origin for the planetesimals that formed Jupiter has been proposed [T. Owen et al., *Nature* 402, 269 (1999)]; and Thommes and colleagues [T. W. Thommes, M. J. Duncan, H. F. Levison, *Nature* 402, 635 (1999)] have proposed the formation of Uranus and Neptune in the Jupiter-Saturn region of the solar system.

25. See N. H. Sleep and K. Zahnle, *J. Geophys. Res.* 103, 28529 (1998) and references therein.

26. G. W. Wetherill, *Astrophys. Space Sci.* 212, 23 (1994); see also G. W. Wetherill, *Icarus* 119, 219 (1996).

27. Reviewed by H. H. Kieffer and A. P. Zent, in *Mars,* H. H. Kieffer et al., Eds. (University of Arizona Press, Tucson, 1992), p. 1180.

28. N. H. Sleep, *J. Geophys. Res.-Planet* 99, 5639 (1994).

29. S. A. Stern and W. B. McKinnon, *Astron. J.,* in press.

30. For the latest on the ever-growing list of extrasolar planets, visit cfa-www.harvard.edu/planets.

31. Special thanks go to Ivan Amato for his help in preparing the final version of this essay. Thanks also go to Steven Dick for historical advice. Any attempt to cover all significant areas and endeavors in an essay of this length would lead to a list that is boring to write and to read. I apologize to planetary science colleagues for the inevitable omissions.

3

Genomics: Journal to the Center of Biology

ERIC S. LANDER AND ROBERT A. WEINBERG

When word came toward the end of 2000 that the quest to sequence the entire human genome was all but completed, it sounded to some like old news, merely the formal recognition of an already accepted, albeit magnificent, achievement. After all, for the past few years, it has been perfectly clear that there was nothing but time in the way of having the complete sequence. In hardly a decade, the tools, techniques, and the approaches and mindsets behind the effort have thoroughly infiltrated the realms of biology, medicine, and biotechnology. In this essay, Eric S. Lander, a key player in the genomic revolution, and Robert A. Weinberg, a master in the field of cancer biology, combine forces to chronicle how the quiet, mid–nineteenth century efforts of a lone monk to study biological heredity have evolved into today's extraordinarily revealing arena of genomic science and technology.

ERIC S. LANDER, who has said "the genetic periodic table will soon be at hand," and ROBERT A. WEINBERG, whose laboratory is credited with identifying the first human oncogene (ras), are members of the Whitehead Institute for Biomedical Research and professors in the Department of Biology at the Massachusetts Institute of Technology in Cambridge, Massachusetts.

Genomics Timeline

1865
Gregor Mendel reports the results of his pea plant experiments, from which he discerned several fundamental laws of heredity. His results, which appeared in an obscure journal article in 1866, were ignored for 34 years.

1882
Walther Flemming publishes his observations of tiny threads—later known as chromosomes—inside salamander larvae cells that appear to be dividing.

1900
Hugo de Vries in the Netherlands, Erich Tschermak von Seysenegg in Austria, and Karl Correns in Germany simultaneously rediscover and verify Mendel's principles of heredity.

1902
Walter Sutton points out connection between chromosomes and Mendel's "factors," thereby expanding the science of genetics from the organismal level to the subcellular level.

Without doubt, the greatest achievement in biology over the past millennium has been the elucidation of the mechanism of heredity. Heredity is surely the strangest of physiological processes: Organisms encapsulate instructions for creating a member of their species in their gametes, these instructions are passed on to a fertilized egg, and then they unfold spontaneously to give rise to offspring. The ancient Greeks puzzled over these remarkable phenomena. Hippocrates imagined that instructional particles were gathered together from throughout the adult body, having been shaped by experience, while Aristotle believed that the instructions were constant and inherent in the gametes. But philosophers could do no more than speculate for the ensuing 2,000 years, because there was no way to probe the physical nature of these instructions.

How the nature of heredity came to be understood over the past 200 years is an extraordinary tale of scientific progress. In dizzying succession, biologists found that the heredity instructions followed specific rules of transmission, resided in the chromosomes contained in the nucleus, were embodied in the molecule DNA, were written in a precise genetic code, and could be read out in their entirety to specify organismic shape and function.

The solution to the problem of heredity turned out to have breathtaking elegance and generality. The instructions for assembling every organism on the planet—slugs and sequoias, peacocks and parasites, whales and wasps—are all specified in DNA sequences that can be translated into digital information and stored in a computer for analysis. As a consequence of this revolution, biology in the twenty-first century is rapidly becoming an information science. Hypotheses will arise as often in silico as in vitro. In this chapter, we recount how this came to pass.

Mendel's Revolution: Transmitting the Instructions

Heredity was the province of philosophers until Anton van Leeuwenhoek's invention of the simple microscope in the seventeenth century. Ironically, early microscopic studies diverted the field; observers convinced themselves that they could see tiny, preformed homunculi ensconced within individual spermatozoa. Preformation obviated the need to store and transmit instructions, but it raised perplexing philosophical questions, such as whether the entire human lineage resided, like nested Russian dolls, in Adam's sperm, and what role Eve played.

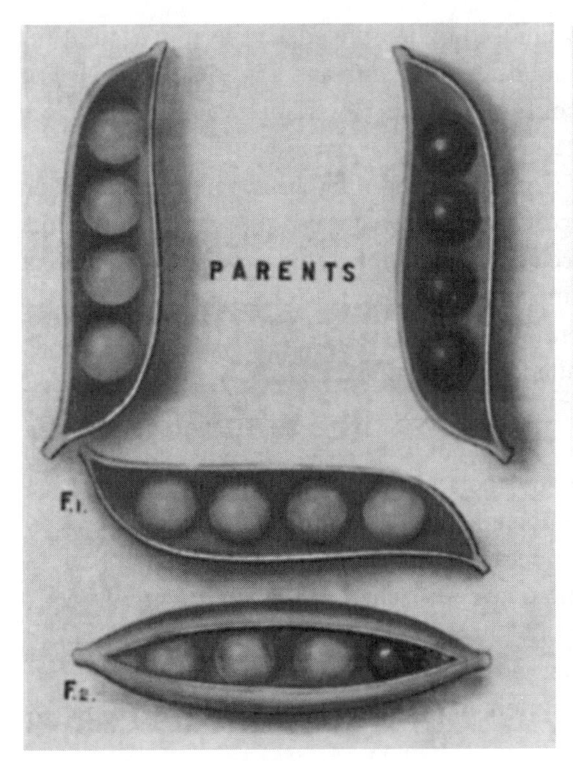

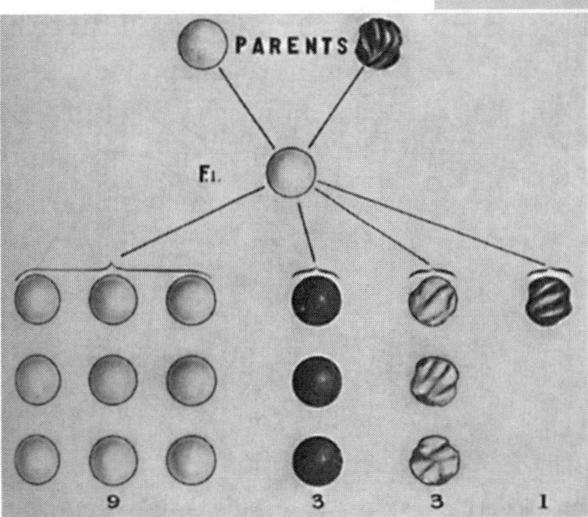

Pea wisdom. Early portraits of heredity laws: segregation of traits (second generation expresses dominant trait; recessive traits reappear in third) and independent assortment of traits (in this case, seed texture and color). *T. H. Morgan, A Critique of the Theory of Evolution (Princeton University Press, Princeton, NJ, 1916).*

1910
Thomas Hunt Morgan and co-workers in the "fly lab" show that some genetically determined traits are sex linked. They also confirm that some trait-determining genes are located on specific chromosomes.

1927
Working with fruit flies, Hermann Muller determines that X rays can cause genetic mutations.

1928
Fred Griffith discovers the phenomenon of transformation, in which some unknown "principle" transforms a harmless strain of bacteria into a virulent one.

Scientific studies of heredity eventually emerged from a more practical quarter—economic forces driving improvements in agriculture. The Age of Discovery from the early fifteenth to the late eighteenth century brought thousands of new plant species to Europe, many of which were propagated and hybridized, improved cultivars being highly prized. The rapidly expanding international trade held great promise for increased economic returns on agricultural products.

One of the hotbeds of interest in agricultural improvement, including animal husbandry, was the city of Brünn (now Brno) in Moravia, center of the textile industry in the Austro-Hungarian Empire of the early nineteenth century. The high price commanded by imported Spanish wool spurred great interest in improving sheep breeds. Breeding programs, however, were conducted largely by trial and error with no underlying rationale. Possessed with remarkable foresight, Brünn's civic leaders organized societies to promote scientific research, citing the importance of discoveries such as those of Copernicus and Newton and expressing hope that the world would someday be similarly indebted to a son of Brünn.

1944

Oswald Avery, Colin MacLeod, and Maclyn McCarty prove that DNA, not protein, embodies the heredity material in most living organisms.

Late 1940s

Erwin Chargaff discovers one-to-one correspondence between adenine and thymine and between cytosine and guanine—a key piece of information for determining the structure of DNA.

1952

Rosalind Franklin obtains X ray diffraction data of DNA, which become central to the elucidation of DNA's molecular structure.

Martha Chase and Alfred Hershey report experiments with bacteriophages that help prove DNA is the molecule of heredity.

1953

James Watson and Francis Crick announce their discovery of the double-helix structure of DNA. They write in a 958-word *Nature* article:

"It has not escaped our notice that the specific pairings we have postulated

This extravagant hope was indeed to be fulfilled. C. F. Napp, head of the Pomological and Oenological Society of Brünn and abbot of the Augustinian monastery, kept his eye out for scientifically trained young men to join his remarkable monastery. The best of these recruits was Gregor Mendel, who had studied physics in Vienna before joining the abbey. The revolution in the understanding of heredity that followed was not triggered by a monk working in isolation who accidentally stumbled upon the laws of genetics. Rather, Mendel worked in an incubator focused on promoting scientific progress in what today would be called agricultural biotechnology.

Out of the garden. Gregor Mendel reported the results of his pea plant experiments, from which he discovered several fundamental laws of heredity, in 1865. His results, which appeared in an obscure journal article in 1866, were ignored for 34 years. *Stock Montage.*

Mendel's pea breeding allowed him to observe genetic dominance and the segregation of traits. In fact, these phenomena had been described qualitatively decades earlier. But Mendel took a quantitative approach, using his physics training and his breeding data to formulate a theory providing, for the first time, a mechanistic description of the laws of heredity.

Mendel proposed that heredity information was passed from parent to offspring in discrete packets, which he called "factors." Different factors were responsible for distinct aspects of a pea plant's appearance, such as seed shape or flower color. His key insight was that the factors occurred in pairs, with one member of the pair being passed on from each parent. The two factors governing a trait might carry conflicting instructions, in which case the voice of one might dominate in determining the appearance of the individual. Nonetheless, the other factor would persist in latent form, and its effects could reappear in later generations in predictable ratios.

Mendel's 1865 report[1] in the *Journal of the Brünn Society of Natural Science* fell on deaf ears. He worked at the periphery of the scientific community, and he published in an obscure journal. But the real problem was that Mendel's formalisms were mathematical and his factors were abstractions. Mendel's laws would only gain a wide audience long after his death, when they could be related to biological realities—visible cellular structures.

Chromosomes: The Cellular Basis of the Instructions

By the mid-1880s, biologists began to recognize that the physical seat of heredity must lie in the cell's nucleus. Microscopists found that recently fertilized eggs carried two equally sized "pronuclei," which later fused. These pronuclei derived from the sperm and the unfertilized egg, which seemed to contribute equally to heredity. Moreover, close examination of a spermatozoon indicated that this cell was little more than a nucleus with a tail.

The most obvious components of the nucleus were its chromosomes, whose behavior could now be studied with precision through greatly improved staining and microscopy techniques. Like the entities that harbored heredity instructions, chromosomes appeared to duplicate with each cycle of cell growth and division. Still, researchers remained unsure about the relation between chromosomes and heredity. Some theories, for example, held that each chromosome carried a complete set of the heredity instructions. The ensuing ferment revived interest in understanding the laws of heredity via breeding experiments.

In the early months of 1900, three researchers—Hugo de Vries, Erich Tschermak von Seysenegg, and Karl Correns—independently reported rediscovering Mendel's work and laws.[2] Their work revealed little more than what Mendel had found thirty-five years earlier, but the scientific community was now primed to listen. The papers sparked the genetics revolution that continued unabated throughout the twentieth century.

An initial challenge was to prove the connection between genes and chromosomes. The most important advances would come from the study of the fruit fly *Drosophila* in the laboratory of Thomas Hunt Morgan at Columbia University. Arguably, the greatest insights came from Alfred Sturtevant, who was performing undergraduate research in Morgan's lab. Sturtevant analyzed

Sex cells. In 1910, Thomas Hunt Morgan and co-workers in the "fly lab" showed that some genetically determined traits are sex linked. They also confirmed that some trait-determining genes are located on specific chromosomes. *AAAS.*

immediately suggest a possible copying mechanism for the genetic material."

Mid-1960s
Marshall Nirenberg, H. Gobind Khorana, and others crack the triplet code that maps messenger RNA codons to specific amino acids.

1969
A team at Harvard Medical School led by Jonathan Beckwith isolates the first gene, specifically, a bacterial gene whose protein product is involved in sugar metabolism.

1970
A team at the University of Wisconsin, led by H. Gobind Khorana, synthesizes a gene from scratch, beginning what might be called chemical genetics.

1972
Using restriction enzymes from Herbert Boyer's research group, Paul Berg and colleagues produce the first recombinant DNA molecules.

1973
The era of genetic engineering begins when Stanley Cohen, Herbert Boyer, and co-workers insert a gene from an African clawed toad into bacterial DNA.

1976
Genentech, the first genetic engineering company, is founded in South San Francisco.

1983
James Gusella and co-workers locate a genetic marker for Huntington's disease on chromosome 4. This leads to scientists having the ability to screen people for a disease without being able to cure it.

Meanwhile, Kary Mullis conceives of the polymerase chain reaction, a chemical DNA replication process that will greatly quicken the pace of genetic science and technology development.

1984
Alec Jeffreys develops "genetic fingerprinting," a molecular biological analog of traditional fingerprinting for identifying individuals by analyzing polymorphic (variable) sequences in their DNA.

a large body of experimental results describing the frequency with which pairs of genes were cotransmitted when passed from parent to offspring. He realized that these data could be explained by a simple model in which the genes were arrayed along a linear "linkage map," with nearby genes being cotransmitted more often than gene pairs located far from one another along his maps. Sturtevant realized that these maps showing the positions of genes must correspond to the threadlike chromosomes.[3] Gene mapping rapidly became a powerful tool of genetics, although the definitive proof of the connection between linkage maps and chromosomes came later in the 1930s with studies by Barbara McClintock on maize chromosomes.[4]

DNA: The Biochemical Basis of the Instructions

The early twentieth century witnessed the birth of another experimental science: biochemistry. This marriage of biology and chemistry sought to understand life by isolating molecules and reconstituting living processes in the nonliving extracts prepared from cells. The biochemists had a clear agenda: to systematically dismantle the notion of vitalism, which held that ineffable "life forces" were responsible for the complex attributes of living cells and tissues. By 1925, they had triumphantly shown that many biochemical reactions could be reproduced in the test tube using the organic catalysts called enzymes. The science of genetics, however, was disconnected from this forward rush of biochemical progress. Genes seemed hopelessly inaccessible: How could one possibly purify heredity in a test tube? Indeed, could heredity ever be understood through biochemistry and the increasing number of molecular species being unearthed in living cells?

A first, tentative foray toward the molecular embodiment of genes came from the 1927 work of Hermann Muller, then in Texas, who demonstrated that X rays could mutate the genes of *Drosophila*.[5] This provided geneticists with a powerful tool. They no longer needed to rely on the spontaneous randomness of nature to generate variants of the "wild-type" genes normally found in flies. Conceptually, Muller's discovery was even more far-reaching, showing that genes were physical entities susceptible to being damaged just like other molecules in the cell. But still, the central question remained: What kind of molecules explained heredity?

A year later, some steps were taken toward an answer. Fred Griffith in England made the serendipitous observation that an extract prepared from virulent, disease-causing *Pneumococcus* bacteria could transmit the trait of virulence to a benign strain. Once the benign bacteria had acquired these instructions, they and their descendants in turn showed all the traits of virulence. The instructions for inducing disease, whatever their nature, persisted in the virulent bacteria long after their death by heat treatment.[6]

By the mid-1930s, Oswald Avery, Colin McLeod, and Maclyn McCarty, working at the Rockefeller Institute in New York, took on the daunting task of purifying the elusive substance that conferred virulence. By 1944 they had the answer. Deoxyribonucleic acid (DNA) molecules extracted from virulent bacteria sufficed to transfer the genetic instructions for virulence. Destruction of the DNA resulted in loss of these instructions, while destruction of bacterial proteins seemed to have no effect on the information transfer.[7]

Their conclusion was controversial. DNA molecules were widely regarded as boring, monotonous chains composed of four nucleotides—ostensibly structural scaffolds of the chromosomes. Protein molecules were far more interesting. They were biochemically and structurally more complex, and for this reason seemed to offer far more possibilities for harboring genetic information. But DNA survived the skeptics. When purified of all but 0.02 percent of contaminating protein, DNA continued to be potent in transmitting genetic information. Most compelling were the 1952 experiments of Alfred Hershey and Martha Chase, who showed that when bacterial viruses inject their genetic information into host cells, DNA enters the cell, while the protein coat remains on the outside.[8]

Still missing was an understanding of how DNA—or any molecule—could store and encode heredity instructions. This intellectual puzzle had attracted the interest of some eclectic physicists, including Niels Bohr and his student Max Delbrück. They struggled to explain the long-term stability of genes in terms of molecules residing in deep potential wells and even suggested that new laws of physics might be needed to explain life. These issues were distilled by Erwin Schrödinger in a brilliant and popular 1945 book, *What is Life?*[9]

Schrödinger proposed that genes must be "aperiodic crystals" consisting of a succession of a small number of isomeric elements whose precise sequence constitutes the heredity code in the manner of the Morse code. Although these ideas did nothing to identify the responsible molecular structures, they did attract many newcomers to the field—including a youthful James Watson, who set off to Cambridge, England, determined to work on the nature of the gene. There, he teamed up with former physicist Francis Crick.

Watson and Crick's revelation of the double-helical structure of DNA struck like a thunderbolt in April 1953.[10] Just as Schrödinger had predicted, DNA was an aperiodic crystal, being composed of four nucleotide bases along

1986

The Human Genome Initiative, later called the Human Genome Project, is announced. The goal is to sequence the entire human genome and provide a complete catalog of every human gene.

1987

A large, collaborative effort yields the first comprehensive human genetic map with 400 signposts.

1988

The National Center for Human Genome Research is created, with the goal of mapping and sequencing all human DNA by 2005.

1990

The Human Genome Project is formally launched, with a completion date set for 2005.

W. French Anderson performs the first gene therapy procedure on a four-year-old girl with an immune disorder known as ADA deficiency. (It didn't work.) Mary-Claire King finds evidence that a gene on chromosome 17 causes an inherited form of breast cancer and increases the risk of ovarian cancer.

its strands. The double helix, by pairing A with T and G with C on opposite strands, explained how genetic information could be copied (the complementarity of nucleotides meant that each strand could serve as a template for the assembly of a complete double helix) and how mutations could arise (occasionally the copying process might go awry). In one stroke, Watson and Crick had explained the key problems of genetics.

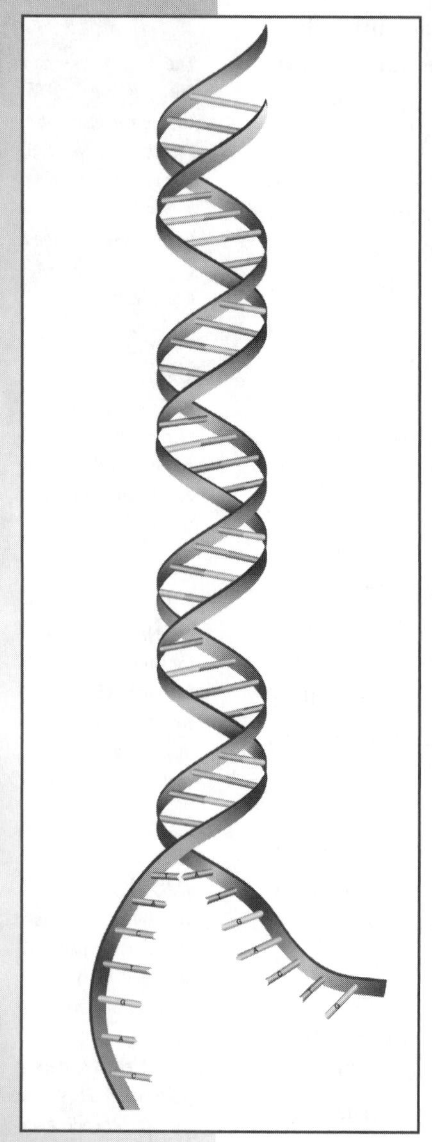

Miracle helices. In 1953, DNA's now iconic molecular structure was first revealed. *AAAS.*

1992

An international collaboration produces a rough map of genetic polymorphism: the variable genetic regions along all 23 pairs of human chromosomes that govern person-to-person biological variation.

The Genetic Code through the Recombinant DNA Revolution: Deciphering the Instructions

The Watson-Crick model made clear that the instructions must be encoded by the sequence of the bases in the strands of the DNA double helix. But how, specifically, were these instructions read out to build the components of a living organism? By 1964, the outlines of the solution had been worked out. The DNA segment corresponding to each gene is copied into a messenger RNA molecule, whose base sequence is then used to direct the synthesis of a specific protein from amino acid building blocks. Marshall Nirenberg used synthetic RNAs to crack the genetic code by which triplets of bases (nucleotides) constitute genetic "words" specifying particular amino acids.[11] In principle, the secret of life had been laid bare.

In practice, there was a catch. Although biologists had deciphered the code for translating DNA information into proteins, they could not yet read any natural DNA sequences—not even the sequence of a single gene out of the thousands present within a cell. They lacked the text on which to practice their newfound deciphering skills. It took another 15 years for this problem to be fully solved by the two recombinant DNA technologies of cloning and sequencing.

Cloning circumvented the limitations of traditional biochemistry, which relies on isolating molecules from a complex mixture based on their chemical idiosyncrasies. The biochemical approach is useless for purifying individual genes because chemically, they are virtually identical—each is simply a stretch of DNA bases. Cloning introduced a new twist on purification: Large genomes were cut into small segments; each was attached to a special "vector" molecule and then introduced into bacterial cells, which faithfully reproduced the foreign DNA as the cells grew. Each bacterium received a single DNA molecule, ensuring that descendant cells would together constitute a "clone," all harboring exact replicas of this specific DNA segment. Scientists thus purified individual DNA segments by propagating them in distinct clones, a collection of which came to be called a "gene library." Experimenters then devised clever steps to enable them to screen the millions of separate clones in a complete gene library and pick out the rare clones carrying the DNA segment, and thus gene, of interest.

The development of the vectors capable of directing the bacterial cell to reproduce the individual DNA segments was a key technical step in the creation of these libraries. Here, biologists exploited highly successful experiments of nature such as viruses and plasmids, which are cellular parasites known to co-opt cells into making hundreds or thousands of copies of viral and plasmid DNA molecules.[12]

DNA sequencing technology formed the other half of the recombinant DNA revolution of the 1970s. Two strategies—one pioneered by Fred Sanger, the other by Walter Gilbert—made it possible to determine with relatively high accuracy the sequences of DNA fragments a few hundred bases long.[13] Soon, individual genes cloned from large cellular genomes became objects of study.

Sequencing technology advanced rapidly, driven by an unquenchable thirst for sequence information. In the late 1970s, an entire doctoral thesis might be devoted to reporting the sequence of a gene of several thousand DNA bases. By century's end, technologists had developed automated sequencing machines capable of cranking out up to a half-million bases per day.

The Genomics Revolution

DNA sequencing soon produced surprises by revealing connections between genes that previously had seemed unrelated. Two early examples involved cancer-causing genes: the oncogenes *sis* and *erbB*. One research team cloned these genes and determined their DNA sequences. Meanwhile, unrelated groups of researchers who were more biochemically inclined isolated two proteins—platelet-derived growth factor (PDGF) and the receptor for epidermal growth factor (EGF)—and determined the amino acid sequences of

1995
The Institute for Genomic Research reports the first complete DNA sequence of the genome of a free-living organism—the bacterium *Haemophilus influenzae.*

1996
The first complete sequence of the genome of a eukaryote (the yeast *Saccharomyces cerevisiae)* is reported by an international effort involving some 600 scientists in Europe, North America, and Japan.

1998
The first genome of a multicellular organism—the 97-megabase DNA sequence of the roundworm *Caenorhabditis elegans*—is published by the *C. elegans* Sequencing Consortium.

2000
Completion of the first draft of the sequence of the entire human genome.

both. To everyone's surprise, the DNA sequences of the oncogenes corresponded nearly perfectly to the amino acid sequences of the well-studied growth-controlling proteins.[14] These identifications revealed instantaneously how the *sis* and *erbB* oncogenes transform normal cells into cancer cells.

Such connections were only the beginning. Comparisons of gene sequences revealed that strikingly similar proteins were encoded in the genomes of organisms as distantly related as yeast and mammals. For example, the proteins governing the progression of a yeast cell through its cycle of growth and division were found in very similar form in the cells of humans. These cross-connections soon numbered in the thousands, then tens of thousands. It became clear that the evolution of life on this planet was stunningly conservative. Once nucleated cells evolved more than 1.5 billion years ago, the great majority of the proteins invented at the time were perpetuated in myriad descendant cells—sometimes with only minor changes. Often the genes encoding these early proteins multiplied and diversified a billion years later, spawning large families of related genes and proteins having diverse, sometimes totally novel functions.

Recognition of gene families produced enormous synergy in research, as the function of one member could often be deduced from that of its known relatives. When the gene responsible for cystic fibrosis was cloned, sequence analysis immediately suggested that it belonged to a family of proteins that transport ions through membranes—a fact that was then readily tested and confirmed in the laboratory.[15]

The connections across vast phylogenetic distances also drove biologists to reconceptualize their research. Those researching organisms such as worms, flies, and yeast began portraying their work as opening windows on the universal rules of life on this planet, not just on the idiosyncrasies of the arcane organisms they studied. Researchers working on sea urchin and frog development found themselves catapulted into cancer research meetings, using a common vocabulary with cancer researchers to describe proteins that play equally important roles in early embryogenesis and in the development of human malignancies.

Sequence analysis also revolutionized the study of evolution, by making it possible to draw phylogenetic trees relating organisms on the basis of similarities in their genes rather than shared physical characteristics. By the 1980s, the availability of vast quantities of sequence data and sophisticated computer-based analytic tools led to a wholesale redrawing of the branches and twigs of the tree of life.

The studies of individual genes represented stunning achievements, but these successes soon promoted an even grander vision: the systematic study of complete genomes, soon referred to as genomics. The first foray into genomics was a proposal to use DNA technology to extend Sturtevant's original concept of genetic mapping to the human being. Instead of tracing the inheritance of visible mutations as had been done in fruit flies, David Botstein and colleagues proposed in 1980 that one could construct a complete genetic map of the human chromosomes by following the inheritance of common DNA sequence variations, termed DNA polymorphisms.[16] Each polymorphism could be used to plant a sequence marker at a specific site on a genetic map of a chromosome. One could then localize genes causing specific human diseases by matching their inheritance patterns with those of the signposts on these genetic maps.

The first success using this strategy came in 1983, when the gene causing Huntington's disease was shown to map to the tip of the short arm of human chromosome 4.[17] The first comprehensive human genetic map with 400 signposts was constructed by 1987, and much denser maps with more than 10,000 such markers were available a decade later. Medical genetics was revolutionized as the genes causing more than 1,000 human diseases were soon mapped to specific chromosomal sites.

An even more expansive vision was expounded in 1985: The entire human genome would be sequenced, providing a complete catalog of every human gene. On its face, the proposal seemed quixotic, a logistical impossibility. The human genome encompasses three billion bases of DNA; then-current sequencing technology could only read out lengths of about 300 bases in each analysis. Decades of work by vast hordes of technicians would surely be required to complete the task.

Moreover, some argued that sequencing the human genome was a fool's errand because the vast majority of it—perhaps 95 percent—does not encode proteins or regulatory information. These sequences were derogatorily labeled "junk DNA." Why, some asked, expend enormous effort to acquire detailed sequence information about DNA that had slim prospect of ever yielding insight into biological function?

But the proposal prevailed. Several years of debate restructured the initial plan into a series of staged subprojects. The relatively small genomes of important experimental organisms—bacteria, yeast, flies, and worms—would be attacked first, before turning to that of the human. Biologically interesting in their own right, these genomes would serve as pilot projects designed to refine the tools for automated sequencing and computational analysis of genomic information. These efforts were organized in 1990 as the International Human Genome Project, biology's first attempt to create a large-scale infrastructure for studying life.

The first project to get under way was the sequencing of the 12-million-base (Mb) genome of the yeast *Saccharomyces cerevisiae,* with sequences of individual chromosomes pouring out between 1992 and 1996 in an international collaboration involving dozens of labs.[18] In 1995, the first complete bacterial genome was produced—the 1.8-Mb *Haemophilus influenzae.* It was all generated in a single laboratory using a "shotgun" technique in which the whole genome is randomly fragmented, the fragments are sequenced, and the results are merged and reassembled into one coherent genome-length sequence.[19]

The results transformed cell biology. For the first time, biologists could enumerate the full complement of genes and proteins required to run a living cell. Included here were the essential hardware components of nucleated (eukaryotic) cells and those of the simpler nonnucleated (prokaryotic) cells.

By 1998, the genome of the first multicellular organism—the 97-Mb DNA sequence of the roundworm *Caenorhabditis elegans*—was published.[20] And sequencing of the genomes of the mustard weed *Arabidopsis thaliana* and the fruit fly *Drosophila melanogaster* was already nearing completion as this past century drew to a close. One truly astounding result, long suspected, was definitively confirmed by these sequencing efforts: The number of distinct genes required to template a complex organism such as the fruit fly (which has some 13,000 genes) was found to be not much greater than the ~6000 carried in the genome of the single-celled baker's yeast.

The pace of sequencing has only quickened. The sequence of the human genome was completed in rough form in 2000 and published in finished form in 2001. Biologists have begun to think of the complete sequence of an organism's genome as a necessary starting point for serious research.

The Future: Global Views of Biology

The availability of the complete parts lists of organisms, that is, catalogs of all their genes and thus all their proteins, has been redirecting biologists toward a global perspective on life processes—to study the role of all genes or all proteins at once. Twentieth-century biology triumphed because of its focus on intensive analysis of the individual components of complex biological systems. The twenty-first century discipline will focus increasingly on the study of entire biological systems, by attempting to understand how component parts collaborate to create a whole. For the first time in a century, reductionists have yielded ground to those trying to gain a holistic view of cells and tissues.

This new approach promises a stunning breadth of perspective. At the same time, it threatens to inundate scientists in a flood of data that will overwhelm their ability to interpret it. Powerful new types of bioinformatics will clearly be required to assimilate and interpret the data that will issue from various types of genomics research. We focus here on a few of these new global perspectives whose outlines are already clear.

Human genetics will be affected in profound ways. Once the human genome is sequenced, a follow-up task will be to understand the spectrum of genetic variation in the human gene pool and its relation to disease. This turns out, surprisingly, to be a tractable problem because of the relatively recent vintage of our species. The current world population of six billion people descends from a few tens of thousands of progenitors who inhabited Africa some 150,000 to 200,000 years ago. Such small populations can maintain only a limited degree of genetic diversity—typically, only a few common variants in the coding sequences of each gene in their genome. Moreover, the few thousand generations of subsequent exponential population expansion have been too few on an evolutionary time scale to alter the spectrum of common variation substantially. As a result, the modern human population has much less intraspecies genetic variation than, for example, chimpanzees. Recent experimental studies have confirmed the limited number of common variants in typical genes, raising the prospect that it will be possible to catalog all of the common variants (alleles) of all human genes.

Such common variants attract enormous interest, because it is suspected that they may hold the key to inborn susceptibility to common diseases. A few cases in point are already known, such as common variants in the apolipoprotein E gene that predispose carriers to Alzheimer's disease or in the clotting factor V gene that predispose carriers to deep venous thrombosis.[21] Some human geneticists believe that such examples represent only the tip of an enormous iceberg. The challenge here will be to identify the full collection of variants and then test their correlation with diseases.

Just as comparison within the human species will be revealing, so too will comparisons between species. Evolution is a grand experiment in which myriad sequence changes within a gene are tried and tested in the crucible of selection. Evolutionary comparison among organisms illuminates those sequences that play important functional roles in protein structure or gene regulation, and thus have been retained unaltered over extended periods of evolutionary time. Identifying evolution's successful experiments will reveal key functional features of important genes and proteins, obviating years of painstaking laboratory experimentation.

Evolutionary sequence comparison should allow us to identify genes that were crucial to the creation of new species; such genes are likely to have undergone strong selection and more rapid sequence evolution. One putative example of such a gene has been proposed in the fruit fly.[22] It would be fascinating to find candidates for the genes and genetic changes that triggered the speciation between our progenitors and those of chimpanzees.

Global approaches are also proving central to efforts to understand the physiology of cells and organisms. Key here will be our ability to survey which genes within a given cell are being read out (expressed) and which ones are silent. A starting point will come from successful monitoring of how the level of each expressed RNA and protein differs among the cells of different tissues in response to different physiologic signals, or in various disease states. Already, microarray detectors exist that allow researchers to measure the RNA levels corresponding to each of the 10,000 or so known genes (still only 30 percent of the total), and various approaches are being developed for studying complex mixtures of expressed proteins—the new science of proteomics.

Because the spectrum of expressed proteins within a cell determines its biology, such comprehensive descriptions will provide the basis for understanding precisely why, for example, brain cells differ from kidney cells. They will identify biological markers characteristic of disease states, leading to techniques for early identification. They will help classify cancers into distinct subtypes, making it possible to know a tumor's lineage, the nature of the genetic mutations that led to its appearance, and, in the long run, whether it will respond to a particular therapy. And they will reveal the strategies of attack that a pathogen launches against its host and the defenses mounted by the host to defeat the invading pathogen.

Proteins that interact physically invariably communicate with one another. So other techniques, such as the "two-hybrid screen" and its relatives, are being developed to identify these interactions. Maps of these associations will, in turn, shed much light on the design of the channels that send and process signals within living cells.

The long-term goal is to use this information to reconstruct the complex molecular circuitry that operates within the cell—to map out the network of interacting proteins that determines the underlying logic of various cellular biological functions including cell proliferation, responses to physiologic stresses, and acquisition and maintenance of tissue-specific differentiation functions. A longer-term goal, whose feasibility remains unclear, is to create mathematical models of these biological circuits and thereby predict these various types of cell biological behavior.

More powerful tools will be needed to realize these goals, at the level of both instrumentation and the computerized processing of biological information, which has now become the cottage industry of bioinformatics. Biologists will require gene-specific reagents to disrupt the function of each component in the cell and study the rippling effects of each such disruption on other genes and proteins within the cell. Various techniques, such as antisense reagents complementary to individual genes and small-molecule screening, are currently being explored with the aim of finding a general technique for disrupting intracellular circuits in a specific, highly targeted fashion. The challenge of disrupting every gene in a human cell is daunting but perhaps not insurmountable—after all, human genes number a mere 30,000 or so, a figure that seems less formidable with the passage of time.

Biology enters this century in possession, for the first time, of the mysterious instruction book first postulated by Hippocrates and Aristotle. How far will this take us in explaining the vast complexity of the biological world? Will we ever be able to draw a protozoan or a peacock, knowing only its DNA sequence? We are hard pressed to provide an answer at the beginning of the century. Still, we proceed with great optimism: The solutions to many problems long resistant to attack are now within reach. The prospects of twenty-first-century biology are surely breathtaking.

At the same time, we must confront this new world soberly and with some trepidation. The genetic diagnostics that can empower patients to seek personalized medical attention may also fuel genetic discrimination. The understanding of the human genetic circuitry that will provide cures for countless diseases may also lead some to conclude that humans are but machines designed to play out DNA cassettes supplied at birth—that the human spirit and human potential are shackled by double-helical chains. So the most serious impact of genomics may well be on how we choose to view ourselves and each another. Meeting these challenges, some quite insidious, will require our constant vigilance, lest we lose sight of why we are here, who we are, and what we wish to become.

References and Notes

1. G. Mendel, *Verh. Naturforsch. Ver. Brünn* 4, 3 (1866). (The original paper was published in *Verh. Naturforsch. Ver. Brünn* 1865, which appeared in 1866.)
2. K. G. Correns, *Ber. dtsch. bot. Ges.* 18, 158 (1900); E. Tschermak von Seysenegg, *Ber. dtsch. bot. Ges.* 18, 232 (1900); H. de Vries, *Ber. dtsch. bot. Ges.* 18, 83 (1900).
3. A. H. Sturtevant, *A History of Genetics* (Harper & Row, New York, 1965); A. H. Sturtevant, *J. Exp. Zool.* 14, 43 (1913).
4. H. B. Creighton and B. McClintock, *Science* 17, 492 (1931).
5. H. J. Muller, *Science* 46, 84 (1927).
6. F. Griffith, *J. Hyg.* 27, 113 (1928).
7. O. T. Avery, C. M. MacLeod, M. McCarty, *J. Exp. Med.* 79, 137 (1944).
8. A. D. Hershey and M. Chase, *J. Gen. Physiol.* 36, 39 (1952).
9. E. Schrödinger, *What is Life?* (Cambridge University Press, New York, 1945).
10. J. D. Watson and F. H. C. Crick, *Nature* 171, 737 (1953).
11. M. W. Nirenberg and J. H. Matthaei, *Proc. Natl. Acad. Sci. U.S.A.* 47, 1588 (1961); M. Nirenberg and P. Leder, *Science* 145, 1399 (1964).
12. J. D. Watson, M. Gilman, J. Witkowski, M. Zoller, *Recombinant DNA* (Freeman, 2nd ed., New York, 1992).
13. A. M. Maxam and W. Gilbert, *Proc. Natl. Acad. Sci. U.S.A.* 74, 560 (1977); F. Sanger, *Science* 214, 1205 (1981).
14. J. Downward et al., *Nature* 307, 521 (1984); R. F. Doolittle et al., *Science* 221, 275 (1983); M. D. Waterfield et al., *Nature* 304, 35 (1983).
15. J. R. Riordan et al., *Science* 245, 1066 (1989).
16. D. Botstein, R. L. White, M. Skolnick, R. W. Davis, *Am. J. Hum. Genet.* 32, 314 (1980).
17. J. F. Gusella et al., *Nature* 306, 234 (1983).
18. R. A. Clayton, O. White, K. A. Ketchum, J. C. Venter, *Nature* 387, 459 (1997).
19. R. D. Fleischmann et al., *Science* 269, 496 (1995).
20. The *C. elegans* Sequencing Consortium, *Science* 282, 2012 (1998).
21. W. J. Strittmatter et al., *Proc. Natl. Acad. Sci. U.S.A.* 90, 1977 (1993); R. M. Bertina et al., *Nature* 369, 64 (1994).
22. C.-T. Ting, S.-C. Tsaur, M.-L. Wu, C.-I. Wu, *Science* 282, 1501 (1998).

23. Related work of E.S.L. is supported by a grant from the National Institutes of Health/National Human Genome Research Institute. The authors would like to thank Maureen T. Murray for her comments on the text.

4

Infectious History

JOSHUA LEDERBERG

Humanity and microbes have been intertwined in the most intimate, rewarding, and tragic ways possible. In this essay, Joshua Lederberg escorts readers on a trans-global journey across this twined history. As he tells what is partly a horror story and partly a celebration of microbial diversity, Lederberg suggests that long-term human survival may require us to embrace a more microbial point of view.

JOSHUA LEDERBERG, a former president of Rockefeller University in New York City, is a Sackler Foundation Scholar heading the university's Laboratory of Molecular Genetics and Information. A geneticist and microbiologist, he received the Nobel Prize in Physiology or Medicine in 1958 for his work in bacterial genetics. He also has worked closely with the National Academies' Institute of Medicine and the Centers for Disease Control and Prevention on analytical and policy issues associated with emerging infections.

Infectious Disease Timeline

1346
Black Death begins spreading in Europe.

1492
Christopher Columbus initiates European-American contact, which leads to transmission of European diseases to the Americas and vice versa.

1530
Girolamo Fracastoro puts forward an early version of the germ theory of disease.

1627
Cinchona bark (quinine) is brought to Europe to treat malaria.

1683
Anton van Leeuwenhoek uses his microscopes to observe tiny animalcules (later known as bacteria) in tooth plaque.

In 1530, to express his ideas on the origin of syphilis, the Italian physician Girolamo Fracastoro penned *Syphilis, sive morbus Gallicus* (Syphilis, or the French disease) in verse. In it he taught that this sexually transmitted disease was spread by "seeds" distributed by intimate contact. In later writings, he expanded this early "contagionist" theory. Besides contagion by personal contact, he described contagion by indirect contact, such as the handling or wearing of clothes, and even contagion at a distance, that is, the spread of disease by something in the air.

Girolamo Fracastoro. *CNRI.*

Fracastoro was anticipating, by nearly 350 years, one of the most important turning points in biological and medical history—the consolidation of the germ theory of disease by Louis Pasteur and Robert Koch in the late 1870s. As we enter the twenty-first century, infectious disease is fated to remain a crucial research challenge, one of conceptual intricacy and of global consequence.

The Incubation of a Scientific Discipline

Many people laid the groundwork for the germ theory. Even the terrified masses touched by the Black Death (bubonic plague) in Europe after 1346 had some intimation of a contagion at work. But they lived within a cognitive framework in which scapegoating, say, of witches and Jews, could more "naturally" account for their woes. Breaking that mindset would take many innovations, including microscopy in the hands of Anton van Leeuwenhoek. In 1683, with one of his new microscopes in hand, he visualized bacteria among the animalcules harvested from his own teeth. That opened the way to visualize some of the dreaded microbial agents eliciting contagious diseases.

Let's get small. Anton van Leeuwenhoek used his microscopes to observe tiny animalcules (later known as bacteria) in tooth plaque in 1683. *Visuals Unlimited.*

There were pre–germ-theory advances in therapy, too. Jesuit missionaries in malaria-ridden Peru had noted the native Indians' use of *Cinchona* bark. In 1627, the Jesuits imported the bark (harboring quinine, its anti-infective ingredient) to Europe for treating malaria. Quinine thereby joined the rarified pharmacopoeia—including opium, digitalis, willow *(Salix)* bark with its analgesic salicylates, and little else—that prior to the modern era afforded patients any benefit beyond placebo.

Beginning in 1796, Edward Jenner took another major therapeutic step—the development of vaccination—after observing that milkmaids exposed to cowpox didn't contract smallpox. He had no theoretical insight into the biological mechanism of resistance to the disease, but vaccination became a lasting prophylactic technique on purely empirical grounds. Jenner's discovery had precursors. "Hair of the dog" is an ancient trope for countering injury and may go back to legends of the emperor Mithridates, who habituated himself to lethal doses of poisons by gradually increasing the dose. We now understand more about a host's immunological response to a cross-reacting virus variant.

Sanitary reforms also helped. Arising out of revulsion over the squalor and stink of urban slums in England and the United States, a hygienic movement tried to scrub up dirt and put an end to sewer stenches. The effort had some health impact in the mid–nineteenth century, but it failed to counter diseases spread by fleas and mosquitoes or by personal contact, and it often even failed to keep sewage and drinking water supplies separated.

It was the germ theory—which is credited to Pasteur (a chemist by training) and Koch (ultimately a German professor of public health)—that set a new course for studying and contending with infectious disease. Over the second half of the nineteenth century, these scientists independently synthesized historical evidence with their own research into the germ theory of disease.

Pasteur helped reveal the vastness of the microbial world and its many practical applications. He found microbes to be behind the fermentation of sugar into alcohol and the souring of milk. He developed a heat treatment (pasteurization, that is) that killed microorganisms in milk, which then no longer transmitted tuberculosis or typhoid. And he too developed new vaccines. One was a veterinary vaccine against anthrax. Another was against rabies and was first used in humans in 1885 to treat a young boy who had been bitten by a rabid dog.

One of Koch's most important advances was procedural. He articulated a set of logical and experimental criteria, later restated as "Koch's Postulates," as

1796
Edward Jenner develops technique of vaccination, at first against smallpox.

1848
Ignaz Semmelweis introduces antiseptic methods.

1854
John Snow recognizes link between the spread of cholera and drinking water supplies.

1860s
Louis Pasteur concludes that infectious diseases are caused by living organisms called "germs." An early practical consequence was Joseph Lister's development of antisepsis by using carbolic acid to disinfect wounds.

1876
Robert Koch validates germ theory of disease and helps initiate the science of bacteriology with a paper pinpointing a bacterium as the cause of anthrax.

1880
Louis Pasteur develops method of attenuating a virulent pathogen (for chicken cholera) so that it immunizes but does not infect; in 1881 he devises an anthrax vaccine and in 1885, a rabies vaccine.

Charles Laveran finds malarial parasites in erythrocytes of infected people and shows that the parasite replicates in the host.

1890
Emil von Behring and Shibasaburo Kitasato discover diphtheria antitoxin serum, the first rational approach to therapy for infectious disease.

1891
Paul Ehrlich proposes that antibodies are responsible for immunity.

1892
The field of virology begins when Dmitri Ivanowski discovers exquisitely small pathogenic agents, later known as viruses, while searching for the cause of tobacco mosaic disease.

1899
Organizing meeting of the Society of American Bacteriologists—later

a standard of proof for researchers' assertions that a particular bacterium caused a particular malady. In 1882, he identified the bacterium that causes tuberculosis; a year later he did the same for cholera. Koch also left a legacy of students (and rivals) who began the systematic search for disease-causing microbes: The golden age of microbiology had begun.

Just as the nineteenth century was ending, the growing world of microbes mushroomed beyond bacteria. In 1892, the Russian microbiologist Dmitri Ivanowski, and in 1898, the Dutch botanist Martinus Beijerinck, discovered exquisitely tiny infectious agents that could pass through bacteria-stopping filters. Too small to be seen with the conventional microscope, these agents were described as "filtrable *[sic]* viruses." In fact, while the focus of the current discussion is on human (and animal) disease, these initial discoveries of viruses, their chemical purification, and many others bearing on host specificity and genetics of disease resistance have come from plant pathology.

With a foundation of germ theory in place even before the twentieth century, the study of infectious disease was ready to enter a new phase. Microbe hunting became institutionalized, and armies of researchers systematically applied scientific analyses to understanding disease processes and developing therapies.

During the early acme of microbe hunting, from about 1880 to 1940, however, microbes were all but ignored by mainstream biologists. Medical microbiology had a life of its own, but it was almost totally divorced from general biological studies. Pasteur and Koch were scarcely mentioned by the founders of cell biology and genetics. Instead, bacteriology was taught as a specialty in medicine, outside the schools of basic zoology and botany. Conversely, bacteriologists scarcely heard of the conceptual revolutions in genetic and evolutionary theory.

Bacteriology's slow acceptance was partly due to the minuscule dimensions of microbes. The microscopes of the nineteenth and early twentieth centuries could not resolve internal microbial anatomy with any detail. Only with the advent of electron microscopy in the 1930s did these structures (nucleoids, ribosomes, cell walls and membranes, flagella) become discernible. Prior to that instrumental breakthrough, most biologists had little, if anything, to do with bacteria and viruses. When they did, they viewed such organisms as mysteriously precellular. It was still an audacious leap for René Dubos to entitle his famous 1945 monograph "The Bacterial Cell."

The early segregation of bacteriology and biology per se hampered the scientific community in recognizing the prospects of conducting genetic investigation with bacteria. So it is ironic that the pivotal discovery of molecular genetics—that genetic information resides in the nucleotide sequence of DNA—arose from studies on serological types of pneumococcus, studies needed to monitor the epidemic spread of pneumonia.

This key discovery was initiated in 1928 by the British physician Frederick Griffith. He found that extracts of a pathogenic strain of pneumococcus could transform a harmless strain into a pathogenic one. The hunt was then on to identify the "transforming factor" in the extracts. In 1944, Oswald Avery, Colin MacLeod, and Maclyn McCarty reported in the *Journal of Experimental Medicine* that DNA was the transforming factor. Within a few years, they and others ruled out skeptics' objections that protein coextracted with the DNA might actually be the transforming factor.

Those findings rekindled interest in what was really going on in the life cycle of bacteria. In particular, they led to my own work in 1946 on sexual conjugation in *Escherichia coli* and to the construction of chromosome maps emulating what had been going on in the study of the genetics of fruit flies, maize, and mice for the prior 45 years. Bacteria and bacterial viruses quickly supplanted fruit flies as the test-bed for many of the subsequent developments of molecular genetics and the biotechnology that followed. Ironically, during this time, we were becoming nonchalant about microbes as etiological agents of disease.

Despite its slow emergence, bacteriology was already having a large impact. Its success is most obviously evidenced by the graying of the population. That public health has been improving—due to many factors, especially our better understanding of infectious agents—is graphically shown by the vital statistics. These began to be diligently recorded in the United States after 1900 in order to guide research and apply it to improving public health. The U.S. experience stands out in charts depicting life expectancy at birth through the century. The average life span lengthened dramatically: from 47 years in 1900 to today's expectation of 77 years (74 years for males and 80 for females).* Similar trends are seen in most other industrialized countries, but the gains have been smaller in economically and socially depressed countries.

Other statistics reveal that the decline in mortality ascribable to infectious disease accounted for almost all of the improvement in longevity up to 1950,

*This sex difference in life expectancy is partly explained by the ability of two X chromosomes to buffer against accumulated recessive mutations and is illustrated by the prevalence in males of color blindness and hemophilia. Another factor is the gender-related difference in self-destructive behaviors.

to be known as the American Society for Microbiology—is held at Yale University.

1900
Based on work by Walter Reed, a commission of researchers shows that yellow fever is caused by a virus from mosquitoes; mosquito-eradication programs are begun.

1905
Fritz Schaudinn and Erich Hoffmann discover bacterial cause of syphilis—*Treponema pallidum.*

1911
Francis Rous reports on a viral etiology of a cancer (Rous sarcoma virus).

1918–19
Epidemic of "Spanish" flu causes at least 25 million deaths.

1928
Frederick Griffith discovers genetic transformation phenomenon in *pneumococci,* thereby establishing a foundation of molecular genetics.

1929
Alexander Fleming reports discovering penicillin in mold.

1935
Gerhard Domagk synthesizes the antimetabolite Prontosil, which kills *Streptococcus* in mice.

1937
Ernst Ruska uses an electron microscope to obtain first pictures of a virus.

1941
Selman Waksman suggests the word "antibiotic" for compounds and preparations that have antimicrobial properties; two years later, he and colleagues discover streptomycin, the first antibiotic effective against tuberculosis, in a soil fungus.

when life expectancy had reached 68. The additional decade of life expectancy for babies born today took the rest of the century to gain. Further improvements now appear to be on an asymptotic trajectory: Each new gain is ever harder to come by, at least pending unpredictable breakthroughs in the biology of aging.

The mortality statistics fluctuated considerably during the first half of the last century. Much of this instability was due to sporadic outbreaks of infections such as typhoid fever, tuberculosis, and scarlet fever, which no longer have much statistical impact. Most outstanding is the spike due to the great influenza pandemic of 1918–19 that killed 25 million people worldwide—comparable to the number of deaths in the Great War. Childhood immunization and other science-based medical interventions have played a significant role in the statistical trends also. So have public health measures, among them protection of food and water supplies, segregation of coughing patients, and personal hygiene. Overall economic growth has also helped by contributing to less crowded housing, improved working conditions (including sick leave), and better nutrition.

As infectious diseases have assumed lower rankings in mortality statistics, other killers—mostly diseases of old age, affluence, and civilization—have moved up the ladder. Heart disease and cancer, for example, have loomed as larger threats over the past few decades. Healthier lifestyles, including less smoking, sparer diets, more exercise, and better hygiene, have been important countermeasures. Prophylactic medications such as aspirin, as well as medical and surgical interventions, have also kept people alive longer.

The 1950s were notable for the "wonder drugs"—the new antibiotics penicillin, streptomycin, chloramphenicol, and a growing list of others that at times promised an end to bacteria-based disease. Viral pathogens have offered fewer routes to remedies, except for vaccines, such as Jonas Salk's and Albert Sabin's polio vaccines. These worked by priming immune systems for later challenges by the infectious agents. Old vaccines, including Jenner's small-

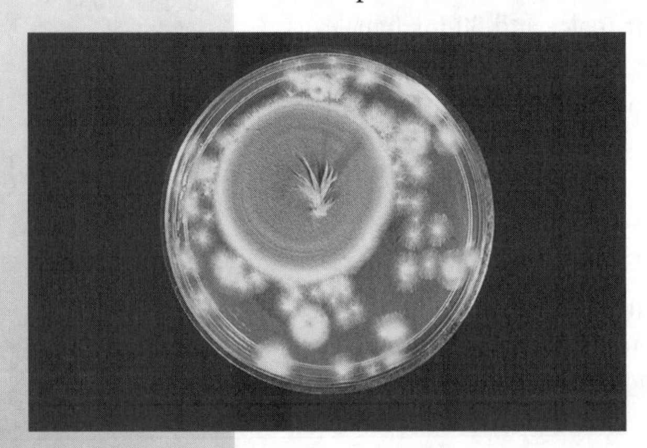

Breaking the mold. In 1929, Alexander Fleming reported discovering penicillin in mold. *P. Barber/CMSP.*

pox vaccine, were also mobilized in massive public health campaigns, sometimes with fantastic results. By the end of the 1970s, smallpox became the first disease to be eradicated from the human experience.

Confidence about medicine's ability to fight infectious disease had grown so high by the mid-1960s that some optimists were portraying infectious microbes as largely conquered. They suggested that researchers shift their attention to constitutional scourges of heart disease, cancer, and psychiatric disorders. These views were reflected in the priorities for research funding and pharmaceutical development. President Nixon's 1971 launch of a national crusade against cancer, which tacitly implied that cancer could be conquered by the bicentennial celebrations of 1976, was an example. Few people now sustain the illusion that audacious medical goals like conquering cancer or infectious disease can be achieved by short-term campaigns.

Wake-Up Calls

The overoptimism and complacency of the 1960s and 1970s was shattered in 1981 with the recognition of AIDS. Since then, the spreading pandemic has overtaken one continent after another with terrible costs. Its spread has been coincident with another wake-up call—the looming problem of antibiotic-resistant microbes. This was a predictable consequence of the evolutionary process operating on microbes challenged by the new selection pressure of antibiotics, arising in part from medical prescriptions and in part from unregulated sales and use in feed for crop animals.

AIDS's causative agent, the human immunodeficiency virus (HIV), is a member of the retrovirus family. These viruses had been laboratory curiosities since 1911, when Francis Peyton Rous discovered the Rous sarcoma virus (RSV) in chickens. Early basic research on retroviruses later helped speed advances in HIV research. By the time AIDS began to spread, RSV had been studied for years as a model for cancer biology, because it could serve as a vector for transferring oncogenes into cells. That work accelerated the characterization of HIV as a retrovirus, and it also helped guide our first steps toward medications that slow HIV infection.

AIDS and HIV have spurred the most concentrated program of biomedical research in history, yet they still defy our counterattacks. And our focus on extirpating the virus may have deflected less ambitious, though more pragmatic, aims, including learning to live with the virus by nurturing in equal measure the immune system that HIV erodes. After all, natural history points

1944
Oswald Avery, Colin MacLeod, and Maclyn McCarty identify DNA as the genetically active material in the pneumococcus transformation.

1946
Edward Tatum and Joshua Lederberg discover "sexual" conjugation in bacteria.

1948
The World Health Organization (WHO) is formed within the U.N.

1952
Renato Dulbecco shows that a single virus particle can produce plaques.

1953
James Watson and Francis Crick reveal the double helical structure of DNA.

Late 1950s
Frank Burnet enunciates clonal selection theory of the immune response.

1960
Arthur Kornberg demonstrates DNA synthesis in cell-free bacterial extract.
François Jacob and Jacques Monod report work on genetic control of enzyme and virus synthesis.

1970
Howard Temin and David Baltimore independently discover that certain RNA viruses use reverse transcription (RNA to reconstitute DNA) as part of their replication cycle.

1975
Asilomar conference sets standards for the containment of possible biohazards from recombinant DNA experiments with microbes.

1979
Smallpox eradication program of WHO is completed; the world is declared free of smallpox.

1981
AIDS first identified as a new infectious disease by U.S. Centers for Disease Control and Prevention.

to analogous infections in simians that have long since achieved a mutually tolerable state of equilibrium.

Costly experiences with AIDS and other infectious agents have led to widespread reexamination of our cohabitation with microbes. Increased monitoring and surveillance by organizations such as the U.S. Centers for Disease Control and Prevention (CDC) and the World Health Organization (WHO) have revealed a stream of outbreaks of exotic diseases. Some have been due to the new importation of microbes (such as cholera in the Southern Hemisphere); some to older parasites (such as *Legionella*) that have been newly recognized as pathogenic; and some to newly evolved antibiotic-resistant pneumonia strains.

Even maladies that had never before been associated with infectious agents recently have been revealed as having microbial bases. Prominent among these are gastric ulcers, which previously had been attributed almost entirely to stress and other psychosomatic causes. Closer study, however, has shown a *Helicobacter* to be the major culprit. Researchers are now directing their speculations away from stress and toward *Chlamydia* infection as a cause of atherosclerosis and coronary disease.

The litany of wake-up calls goes on. Four million Americans are estimated to be infected with the hepatitis C virus (HCV), mainly by transfusion of contaminated blood products from an era prior to contemporary vigilance. The HCV continues to be disseminated via self-inoculation with contaminated needles. This population now is at significant risk for developing liver cancer. Those harboring hepatitis C must be warned to avoid alcohol and other hepatotoxins, and they must not donate blood.

Smaller but lethal outbreaks of dramatic, hypervirulent viruses have been raising public fear. Among these are the Ebola virus outbreak in Africa in 1976 and again in 1995 and the hantavirus outbreak in the U.S. Southwest in 1993. In hindsight, these posed less of a public health risk than the publicity they received might have suggested. Still, studying them and uncovering ecological factors that favor or thwart their proliferation is imperative because of their potential to mutate into more diffusible forms.

Our vigilance is mandated also by the facts of life: The processes of gene reassortment in flu viruses, which are poorly confined to their canonical hosts (birds, swine, and people), goes on relentlessly and is sure to regenerate human-lethal variants. Those thoughts were central in 1997 when the avian flu H5N1 transferred into a score of Hong Kong citizens, a third of whom died. It is likely that the resolute actions of the Hong Kong health authorities, which destroyed two million chickens, stemmed that outbreak and averted the possibility of a worldwide spread of H5N1.

Complacency is not an option in these cases, as other vectors, including wildfowl, could become carriers. In Malaysia, a new infectious entity, the Nipah virus, killed up to 100 people in 1999; authorities there killed a million livestock to help contain the outbreak. New York had a smaller-scale scare in the summer of 1999 with the unprecedented appearance of bird- and mosquito-borne West Nile encephalitis, although the mortality rate was only a few percent of those infected. We need not wonder whether we will see outbreaks like these again. The only questions are when and where?

These multiple wake-up calls to the infectious disease problem have left marks in vital statistics. From midcentury to 1982, the U.S. mortality index (annual deaths per 100,000) attributable to infection had been steady at about 30. But from 1982 to 1994, the rate doubled to 60. (Keep in mind that the index was 500 in 1900 and up to 850 in 1918–19 due to the Spanish flu epidemic.) About half of the recent rise in deaths is attributable to AIDS; much of the rest is due to respiratory disease, antibiotic resistance, and hospital-acquired infection.

The Microbial World Wide Web

The field of molecular genetics, which began in 1944 when DNA was proven to be the molecule of heredity in bacteria-based experiments, ushered microbes into the center of many biological investigations. Microbial systems now provide our most convenient models for experimental evolution. Diverse mechanisms for genetic variation and recombination uncovered in such systems are spelled out in ponderous monographs. Assays for chemical mutagenesis (e.g., the Ames test using Salmonella) are now routinely carried out on bacteria, because microbial DNA is so accessible to environmental insult. Mutators (genes that enhance variability) abound and may be switched on and off by different environmental factors. The germs' ability to transfer their own genetic scripts, via processes such as plasmid transfer, means they can exchange biological innovations including resistance to antibiotics.

Indeed, the microbial biosphere can be thought of as a World

1982
Stanley Prusiner finds evidence that a class of infectious proteins, which he calls prions, cause scrapie in sheep.

1983
Luc Montagnier and Robert Gallo announce their discovery of the human immunodeficiency virus that is believed to cause AIDS.

1984
Barry Marshall shows that isolates from ulcer patients contain the bacterium later known as *Helicobacter pylori.* The discovery ultimately leads to a new pathogen-based etiology of ulcers.

1985
Robert Gallo, Dani Bolognesi, Sam Broder, and others show that AZT inhibits HIV action in vitro.

1988
Kary Mullis reports basis of polymerase chain reaction (PCR) for detection of even single DNA molecules.

1995
J. Craig Venter, Hamilton Smith, Claire Fraser, and colleagues at The Institute for Genomic Research elucidate the first complete genome sequence of a microorganism: *Haemophilus influenzae.*

1996
Implied link between bovine spongiform encephalopathy ("mad cow disease") and human disease syndrome leads to large-scale controls on British cattle.

1999
New York City experiences outbreak of West Nile encephalopathy transmitted by birds and mosquitoes.

c. 2000
Antibiotic-resistant pathogens are spreading in many environments.

For more extensive chronological listings, see "Microbiology's fifty most significant events during the past 125 years," poster supplement to *ASM News* 65(5), 1999.

Wide Web of informational exchange, with DNA serving as the packets of data going every which way. The analogy isn't entirely superficial. Many viruses can integrate (download) their own DNA into host genomes, which subsequently can be copied and passed on: Hundreds of segments of human DNA originated from historical encounters with retroviruses whose genetic information became integrated into our own genomes.

What makes microbial evolution particularly intriguing, and worrisome, is a combination of vast populations and intense fluctuations in those populations. It's a formula for top-speed evolution. Microbial populations may fluctuate by factors of 10 billion on a daily cycle as they move between hosts, or as they encounter antibiotics, antibodies, or other natural hazards. A simple comparison of the pace of evolution between microbes and their multicellular hosts suggests a millionfold or billionfold advantage to the microbe. A year in the life of bacteria would easily match the span of mammalian evolution!

By that metric, we would seem to be playing out of our evolutionary league. Indeed, there's evidence of sporadic species extinctions in natural history, and our own human history has been punctuated by catastrophic plagues. Yet we are still here! Maintaining that status within new contexts in which germs and hosts interact in new ways almost certainly will require us to bring ever more sophisticated technical wit and social intelligence to the contest.

—J.L.

Our Wits Versus their Genes

As our awareness of the microbial environment has intensified, important questions have emerged. What puts us at risk? What precautions can and should we be taking? Are we more or less vulnerable to infectious agents today than in the past? What are the origins of pathogenesis? And how can we use deeper knowledge to develop better medical and public health strategies? Conversely, how much more can the natural history of disease teach us about fundamental biological and evolutionary mechanisms?

An axiomatic starting point for further progress is the simple recognition that humans, animals, plants, and microbes are cohabitants of the planet. That leads to refined questions that focus on the origin and dynamics of instabilities within this context of cohabitation. These instabilities arise from two main sources loosely definable as ecological and evolutionary.

Ecological instabilities arise from the ways we alter the physical and biological environment, the microbial and animal tenants (humans included) of these environments, and our interactions (including hygienic and therapeutic interventions) with the parasites. The future of humanity and microbes likely will unfold as episodes of a suspense thriller that could be titled *Our Wits Versus their Genes.*

We already have used our wits to increase longevity and lessen mortality. That simultaneously has introduced irrevocable changes in our demographics and our own human ecology. Increased longevity, economic productivity, and other factors have abetted a global population explosion from about 1.6 billion in 1900 to its present level above six billion. That same population increase has fostered new vulnerabilities: crowding of humans, with slums cheek by jowl with jet-setters' villas; the destruction of forests for agriculture and suburbanization, which has led to closer human contact with disease-carrying rodents and ticks; and routine long-distance travel.

Travel around the world can be completed in less than 80 hours (compared to the 80 days of Jules Verne's nineteenth-century fantasy), constituting a historic new experience. This long-distance travel has become quotidian: Well over a million passengers, each one a potential carrier of pathogens, travel daily by aircraft to international destinations. International commerce, especially in foodstuffs, only adds to the global traffic of potential pathogens and vectors. Because the transit times of people and goods now are so short compared to the incubation times of disease, carriers of disease can arrive at their destination before the danger they harbor is detectable, reducing health quarantine to a near absurdity.

Our systems for monitoring and diagnosing exotic diseases have hardly kept pace with this qualitative transformation of global human and material exchange. This new era of global travel will redistribute and mix people, their cultures, their prior immunities, and their inherited predispositions, along with pathogens that may have been quiescent at other locales for centuries.

This is not completely novel, of course. The most evident precedent unfolded during the European conquest of America, which was tragically abetted by pandemics of smallpox and measles imported into native populations by the invading armies. In exchange, Europeans picked up syphilis's *Treponema,* in which Fracastoro discerned contagion at work.

Medical defense against the interchange of infectious disease did not exist in the sixteenth century. In the twenty-first century, however, new medical technologies will be key parts of an armamentarium that reinforces our own immunological defenses. This dependence on technology is beginning to be recognized at high levels of national and international policy-making. With the portent of nearly instant global transmission of pathogenic agents, it is ever more important to work with international organizations like WHO for global health improvement. After all, the spread of AIDS in America and Europe in the 1980s and 1990s was due, in part, to an earlier phase of near obliviousness to the frightful health conditions in Africa. One harbinger of the kind of high-tech wit we will need for defending against outbreaks of infectious disease is the use of cutting-edge communications technology and the Internet, which already have been harnessed to post prompt global alerts of emerging diseases (see osi.oracle.com:8080/promed/promed.home).

Moving Targets

"Germs" have long been recognized as living entities, but the realization that they must inexorably be evolving and changing has been slow to sink in to the ideology and practice of the public health sector. This lag has early roots. In the nineteenth century, Koch was convinced that rigorous experiments would support the doctrine of monomorphism: that each disease was caused by a single invariant microbial species rather than by the many that often showed up in culture. He argued that most purported "variants" were probably alien bacteria that had floated into the petri dishes from the atmosphere.

Koch's rigor was an essential riposte to careless claims of interconvertibility—for example, that yeasts could be converted into bacteria. It also helped untangle confusing claims of complex morphogenesis and life cycles among common bacteria. But strict monomorphism was too rigid, and even Koch eventually relented, admitting the possibility of some intrinsic variation rather than contamination. Still, for him and his contemporaries, variation remained a phenomenological and experimental nuisance rather than the essence of microbes' competence as pathogens. The multitude of isolable species was confusing enough to the epidemic tracker; it would have been almost too much to bear to have to cope with constantly emerging variants with altered serological specificity, host affinity, or virulence.

Even today it would be near heresy to balk at the identification of the great plague of the fourteenth century with today's *Yersinia pestis;* but we cannot readily account for its pneumonic transmission without guessing at some intrinsic adaptation at the time to aerosol conveyance. Exhumations of ancient remains might still furnish DNA evidence to test such ideas.

We now know and accept that evolutionary processes elicit changes in the genotypes of germs and of their hosts. The idea that infection might play an important role in natural selection sank in after 1949 when John B. S. Haldane conjectured that the prevalence of hemoglobin disorders in Mediterranean peoples might be a defense against malaria. That idea developed into the first concrete example of a hereditary adaptation to infectious disease.

Haldane's theory preceded Anthony C. Allison's report of the protective effect of heterozygous hemoglobinopathy against falciparum malaria in Africa. The side effects of this bit of natural genetic engineering are well known: When this beneficial polymorphism is driven to higher gene frequencies, the homozygous variant becomes more prevalent and with it the heavy human and societal burden of sickle cell disease.

We now have a handful of illustrations of the connection between infection and evolution. Most are connected to malaria and tuberculosis, which are so prevalent that genetic adaptations capable of checking them have been strongly selected. The same prevalence also makes their associated adaptations more obvious to researchers. A newly reported link between infection and evolution is the effect of a *ccr5* (chemokine receptor) deletion, a genetic alteration that affords some protection against AIDS. It would be interesting to know what factors—another pathogen perhaps—may have driven that polymorphism in earlier human history.

One lesson to be gleaned from this coevolutionary dynamic is how fitful and sporadic human evolution is when our slow and plodding genetic change is pitted against the far more rapidly changing genomes of microbial pathogens.

We have inherited a robust immune system, but little has changed since its early vertebrate origins 200 million years ago. In its inner workings, immunity is a Darwinian struggle: a randomly generated diversification of leukocytes that collectively are prepared to duel with a lifetime of unpredictable invaders. But these duels take place in the host soma; successful immunological encounters do not become genetically inscribed and passed on to future generations of the host. By contrast, the germs that win the battles quickly proliferate their successful genes, and they can use those enhancements to go on to new hosts, at least in the short run.

The human race evidently has withstood the pathogenic challenges encountered so far, albeit with episodes of incalculable tragedy. But the rules of encounter and engagement have been changing; the same record of survival may not necessarily hold for the future. If our collective immune systems fail to keep pace with microbial innovations in the altered contexts we have created, we will have to rely still more on our wits.

Evolving Metaphors of Infection: Teach War No More

New strategies and tactics for countering pathogens will be uncovered by finding and exploiting innovations that evolved within other species in defense against infection. But our most sophisticated leap would be to drop the manichaean view of microbes—"We good; they evil." Microbes indeed have a knack for making us ill, killing us, and even recycling our remains to the geosphere. But in the long run microbes have a shared interest in their hosts' survival: A dead host is a dead end for most invaders, too. Domesticating the host is the better long-term strategy for pathogens.

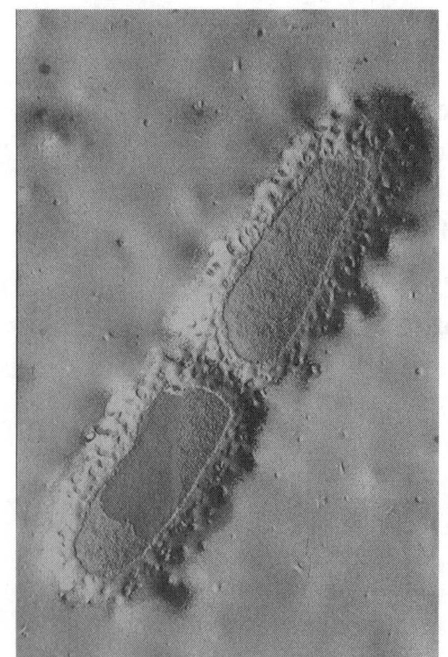

Home team. Escherichia coli are among humans' lifelong symbionts. *J. L. Carson/CMSP.*

We should think of each host and its parasites as a superorganism with the respective genomes yoked into a chimera of sorts. The power of this sociological development could not be more persuasively illustrated than by the case of mitochondria, the most successful of all microbes. They reside inside every eukaryote cell (from yeast to protozoa to multicellular organisms), in which they provide the machinery of oxidative metabolism. Other bacteria have taken similar routes into plant cells and evolved there into chloroplasts—the primary harvesters of solar energy, which drive the production of oxygen and the fixed carbon that nourishes the rest of the biosphere.

These cases reveal how far collaboration between hosts and infecting microbes can go. In the short run, however, the infected host is in fact at metastable equilibrium: The balance could tip toward favorable or catastrophic outcomes.

On the bad side, the host's immune response may be excessive, with autoimmune injuries as side effects. Microbial zeal also can be self-defeating.

As with rogue cancer cells, deviant microbial cells (such as aggressive variants from a gentler parent population) may overtake and kill the host, thereby fomenting their own demise and that of the parent population.

Most successful parasites travel a middle path. It helps for them to have aggressive means of entering the body surfaces and radiating some local toxicity to counter the hosts' defenses, but once established they also do themselves (and their hosts) well by moderating their virulence.

Better understanding of this balancing act awaits further research. And that may take a shift in priorities. For one, research has focused on hypervirulence. Studies into the physiology of homeostatic balance in the infected host qua superorganism have lagged. Yet the latter studies may be even more revealing, as the burden of mutualistic adaptation falls largely on the shoulders of the parasite, not the host. This lopsided responsibility follows from the vastly different evolutionary paces of the two. But then we have our wits, it is to be hoped, for drafting the last word.

To that end, we also need more sophisticated experimental models of infection, which today are largely based on contrived zoonoses (the migration of a parasite from its traditional host into another species). The test organism is usually a mouse, and the procedure is intended to mimic the human disease process. Instead, it is often a caricature.

Injected with a few bugs, the mouse goes belly up the next day. This is superb for in vivo testing of an antibiotic, but it bears little relation to the dynamics of everyday human disease.

Natural zoonoses also can have many different outcomes. In most cases, there will be no infection at all or only mild ones such as the gut ache caused by many *Salmonella enteritidis* species. Those relatively few infectious agents that cause serious sickness or death are actually maladapted to their hosts, to which they may have only recently gained access through some genetic, environmental, or sociological change. These devastatingly virulent zoonoses include psittacosis, Q fever, rickettsiosis, and hantavirus. Partly through lack of prior coevolutionary development with the new host, normal restraints fail.

I suggest that a successful parasite (one that will be able to remain infectious for a long time) tends to display just those epitopes (antigen fragments that stimulate the immune system) that will provoke host responses that a) moderate but do not extinguish the primary infection, and b) inhibit other infections by competing strains of the same species or of other species. According to this speculative framework, the symptoms of influenza evolved as they have in part to ward off other viral infections.

Research into infectious diseases, including tuberculosis, schistosomiasis, and even AIDS, is providing evidence for this view. So are studies of *Helicobacter,* which has been found to secrete antibacterial peptides that inhibit other enteric infections. We need also to look more closely at earlier stages of chronic infection and search for cross-protective factors by which microbes engage one another. HIV, for one, ultimately fails from the microbial perspective when opportunistic infections supervene to kill its host. That result, which is tragic from the human point of view, is a byproduct of the virus's protracted duel with the host's cellular immune system. The HIV envelope and those of related viruses also produce antimicrobials, although their significance for the natural history of disease remains unknown.

Now genomics is entering the picture. Within the past decade, the genomes of many microbes have been completely sequenced. New evidence for the web of genetic interchange is permeating the evolutionary charts. The functional analyses of innumerable genes now emerging are an unexplored mine of new therapeutic targets. It has already shown many intricate intertwinings of hosts' and parasites' physiological pathways. Together with wiser insight into the ground rules of pathogenic evolution, we are developing a versatile platform for developing new responses to infectious disease. Many new vaccines, antibiotics, and immune modulators will emerge from the growing wealth of genomic data.

The lessons of HIV and other emerging infections also have begun taking hold in government and in commercial circles, where the market opportunities these threats offer have invigorated the biotechnology industry. If we do the hard work and never take success for granted (as we did for a while during the last century), we may be able to preempt infectious disasters such as the influenza outbreak of 1918–19 and the more recent and ongoing HIV pandemic.

Perhaps one of the most important changes we can make is to supercede the twentieth-century metaphor of war for describing the relationship between people and infectious agents. A more ecologically informed metaphor, which includes the germs'-eye view of infection, might be more fruitful. Consider that microbes occupy all of our body surfaces. Besides the disease-engendering colonizers of our skin, gut, and mucous membranes, we are host to a poorly cataloged ensemble of symbionts to which we pay scant attention. Yet they are equally part of the superorganism genome with which we engage the rest of the biosphere.

The protective role of our own microbial flora is attested to by the superinfections that often attend specific antibiotic therapy: The temporary decimation of our home-team microbes provides entrée for competitors. Understanding these phenomena affords openings for our advantage, akin to the ultimate exploitation by Dubos and Selman Waksman of intermicrobial competition in the soil for seeking early antibiotics. Research into the microbial ecology of our own bodies will undoubtedly yield similar fruit.

Replacing the war metaphor with an ecological one may bear on other important issues, including debates about eradicating pathogens such as smallpox and polio. Without a clear strategy for sustaining some level of immunity, it makes sense to maintain lab stocks of these and related agents to guard against possible recrudescence. An ecological perspective also suggests other ways of achieving lasting security. For example, domestication of commensal microbes that bear relevant cross-reacting epitopes could afford the same protection as vaccines based on the virulent forms. There might even be a nutraceutical angle: These commensal epitopes could be offered as optional genetically engineered food additives, clearly labeled and meticulously studied.

Another relevant issue that can be recast in an ecological model is the rise in popularity of antibacterial products. This is driven by the popular idea that a superhygienic environment is better than one with germs—the "enemy" in the war metaphor. But too much antibacterial zeal could wipe out the very immunogenic stimulation that has enabled us to cohabit with microbes in the first place.

To help dramatize the interplay of our entourage of associated microbiota with our own physiology, I have suggested that we think of the "microbiome" as a partner to our own "genome." These together, microbiome and genome—and let us not forget the "chondriome," the DNA of our mitochondria-constitute a super-organism, a functional community that is brimming with cellular competition in all of its parts. It is also the de facto unit of natural selection, of competition with the other superorganisms, of the same and of vastly different species, that comprise the terrestrial biosphere.

Ironically, even as I advocate this shift from a war metaphor to an ecology metaphor, war in its historic sense is making that more difficult. The darker corner of microbiological research is the abyss of maliciously designed biological warfare (BW) agents and systems to deliver them. What a nightmare for the next millennium! What's worse, for the near future, technology is likely to favor offensive BW weaponry, because defenses will have to cope with a broad range of microbial threats that can be collected today or designed tomorrow.

As a measure of social intelligence and policy, we should push for enforcement of the 1975 BW disarmament convention. The treaty forbids the development, production, stockpiling, and use of biological weapons under any circumstances. One of its articles also provides for the international sharing of biotechnology for peaceful purposes. The scientific and humanistic rationale is self-evident: to enhance and apply scientific knowledge to manage infectious disease, naturally occurring or otherwise.

Further Readings

W. Bulloch, *History of Bacteriology* (Oxford University Press, London, 1938).

R. Dubos, *Mirage of Health: Utopias, Progress, and Biological Change* (Rutgers University Press, New Brunswick, New Jersey, 1987).

J. Lederberg, R. E. Shope, S. C. Oaks Jr., Eds., *Emerging Infections: Microbial Threats to Health in the United States* (National Academy Press, Washington, D.C., 1992, see www.nap.edu/books/0309047412/html/index.html).

G. Rosen, *A History of Public Health* (Johns Hopkins University Press, Baltimore, 1993).

T. D. Brock, *The Emergence of Bacterial Genetics* (Cold Spring Harbor Laboratory Press, Cold Spring Harbor, New York, 1990).

S. S. Morse, Ed., *Emerging Viruses* (Oxford University Press, New York, 1993).

Journal of the American Medical Association theme issue on emerging infections, August 1996.

W. K. Joklik et al., Eds., *Microbiology, a Centenary Perspective* (ASM Press, Herndon, Virginia, 1999).

S. C. Stearns, Ed., *Evolution in Health and Disease* (Oxford University Press, Oxford, New York, 1999).

P. Ewald, "Evolution of Infectious Disease," in *Encyclopedia of Microbiology*, J. Lederberg, Ed. (Academic Press, Orlando, Florida, 2000).

G. L. Mandell, J. E. Bennett, R. Dolin, *Principles and Practice of Infectious Diseases* (Churchill Livingstone, 5th ed., Philadelphia, 2000).

Notable Web Sites

www.lib.uiowa.edu/hardin/md/micro.html

www.idsociety.org

www.asmusa.org

osi.oracle.com:8080/promed

www.cdc.gov/ncidod/EID

5

Designing a New Material World

GREGORY B. OLSON

Underlying the history of technology is the ever-enlarging ability of people to transform the raw substances available in the environment into more useful and capable materials. In this essay, Gregory B. Olson traces some of the pathways that converge into today's multidisciplinary field of materials science and engineering. He contends that the emerging "systems approach," in which research teams approach materials from all angles, including their underlying atomic arrangements and the challenges of commercializing end products, is beginning to transcend traditional empirical discovery of materials. In its place is coming, in Olson's words, an "Age of Design," which will be marked by the systematic invention of unprecedented materials made to order. In this way, materials scientists will tap far more of the expansive well of secrets harbored within the periodic table of the chemical elements.

GREGORY B. OLSON is Wilson-Cook Professor of Engineering Design in the McCormick School of Engineering and a professor in the department of materials science and engineering at Northwestern University in Evanston, Illinois. He is also a founder of the materials design firm, QuesTek Innovations, which is based in Evanston.

Materials Science Timeline

1556
Georgius Agricola's *De re metallica,* a compendium of sixteenth-century mining, metallurgical, and general materials production, is published.

1664
Cartesian corpuscular philosophy recognizes material properties as emerging from a multilevel structure.

1665
Robert Hooke publishes *Micrographia,* which reveals levels of material microstructure never before seen.

1722
René de Réaumur publishes the first technical treatise on iron.

1782
Josiah Wedgwood develops an early form of process control with his invention of the pyrometer for measuring furnace temperatures.

Materials have paced the evolution of technology for millennia. Their importance in the advance of human civilization is apparent in the naming of historical epochs, from the Stone Age through the Bronze and Iron Ages and into the ongoing Silicon Age. The origin of diversity in the material world remains largely mysterious to the public, yet the specialists' ability to understand and manipulate the microstructures of materials has grown explosively over the past half-century. As the new millennium unfolds, a confluence of natural philosophies—one that combines reductionist and synthetic thinking—is ushering in an Age of Design marked by new materials and ways of creating them that go beyond the dreams of the medieval alchemists.

Materials as Systems

The modern view of materials structure was best expressed by the late philosopher-scientist Cyril Stanley Smith.[1] He described a universal multilevel nature of structure with strong interactions among levels and an inevitable interplay of perfection and imperfection at all levels. Smith argued that the materials scientist's distinct view of structure is defined by the desire to understand the structure and property relations underlying the technological and economic value of materials.

This view of matter integrates science and engineering and is built on a natural philosophy that is older than science. Smith identifies the origin of this philosophy within medieval alchemy, whose practitioners are mostly remembered for their attempts to "transmute" base metals into gold. They knew nothing about protons and electrons. Still, Smith asserts that their aspirations, motivations, and even concepts resembled the modern materials engineer's use of complex processing to "transmute" the multilevel microstructure of materials (from the atomic to macroscopic levels) to achieve the essential property of gold, namely, economic value.

The property-driven view of structure and processing for the creation of value, shared by modern materials science and alchemy, embraces an essential complexity of material structure. During the seventeenth-century birth of modern science, Smith identifies a short-lived Golden Age of materials science under the leadership of René Descartes, whose "corpuscular" philosophy inspired the development of a multilevel view of structure to account for the diverse properties of materials.

The prescient grasp of materials achieved by the corpuscularians is well

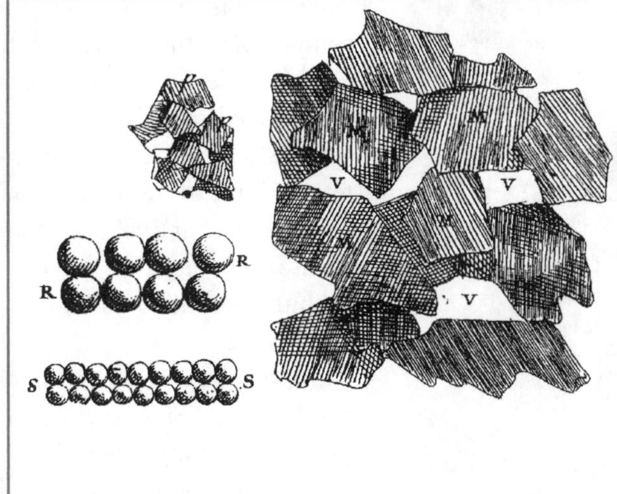

Materials savant. René de Réaumur and his 1722 sketch of steel's anatomy. *Wellcome Institute Library.*

represented by René Antoine Ferchault de Réaumur's 1722 sketch of the structure of quench-hardened steel (see figure). He proposed that a single grain of steel (G), if enlarged, would reveal a set of "molecules" (M) and voids (v). Higher magnification would reveal a substructure of the molecules (pp); and yet higher magnification would show an aperiodic arrangement of spheres. The finest scale Réaumur envisioned corresponds to a periodic packing of spheres, what we might think of as the nanoscale.

This elegant view contained many structural elements of modern materials science. There was no instrumentation with which to validate it, however, so the complex structural view was supplanted by more intellectually compelling but overly simplistic notions. One was Isaac Newton's continuum, which erased structural considerations entirely. The other was John Dalton's atomism, which held that there was only one level of structure that mattered. These conceptual idealizations were sufficiently compelling to divert the corpuscularian framework for two centuries.

We have had the past century to reinvent materials science. The atom and the continuum remain dominant philosophical forces; they are the foundation of existing theoretical tools. But armed with the growing reservoir of structural facts gleaned from instrumentation and computation, the chal-

1808
John Dalton publishes his *New System of Chemical Philosophy,* which establishes atomic theory.

1824
Joseph Aspdin invents portland cement, which remains one of the most used materials in the world.

1839
Charles Goodyear accidentally discovers vulcanization, which ultimately renders raw rubber latex into a widely useful material.

1856
Henry Bessemer patents a process for large-scale steel production.

1860s
Henry Sorby applies light microscopy to the study of the microstructure of metals and rocks.

1869
John Hyatt successfully commercializes celluloid, an artificial plastic material.

1869 and 1870
Dmitri Mendeleev and Julius Lothar Meyer publish versions of what will become known as the Periodic Table of the Chemical Elements.

1886
Charles Hall and Paul Héroult independently discover cost-effective methods for producing aluminum metal from ore.

1893
Floris Osmond discovers martensitic transformation.

1900
Max Planck formulates the idea of quanta, thereby setting the stage for the development of quantum mechanics.

lenge now is to adapt these theoretical tools to understand and control the complex structures of real materials.

The modern view of material structure differs from Réaumur's mainly in the detailed morphologies characteristic at the different length scales of a material's hierarchical structure. In the case of steel, the most significant difference is the overestimation of porosity in the eighteenth-century depictions. And if Réaumur's voids are reinterpreted as "free volume," his sketch becomes a reasonable model of polymeric materials. It shows the remarkable ability of the human mind to infer necessary structure from the contemplation of properties alone.

Smith made another important historical observation about materials development.[2] Since prehistory, people have put newly discovered materials to practical use long before they understood much about them.

Consider the pattern-welded sword blades made and used by Merovingian Franks and Vikings (see figure) as early as the sixth century A.D. By hammer-welding steels of differing carbon content (a technical feature unknown to the early swordsmiths), a laminate composite was fabricated by labor-intensive "hand lay-up" to produce a hybrid structure with a tough core supporting a hard cutting edge.

The actual mechanism of the quench hardening of steel, which is responsible for the hardened sword edge, was hotly debated even in the early twentieth century. There were two factions: the "allotropists," who favored structural transformation, and the "carbonists," who pegged the hardening mechanism on dissolved carbon. A crucial observation was the discovery by Floris Osmond in 1893 of "martensite" in steel, the microstructural form that iron

Cutting edge. Etched and repolished tip region of sixth-century sword. *G. B. Olson.*

assumes during the quenching process. We now know that the combination of a martensitic transformation and the redistribution of trapped interstitial carbon (along with ambient aging during which additional microstructural evolution occurs) underlies the edge hardness of the sixth-century swords.

The ancient swords embody another illustration of how art and craft has traditionally preceded science. The swordsmiths' use of chemical etching to bring out aesthetic metallic patterns set the foundation for modern metallographic observation of microstructure established by Henry Sorby in the nineteenth century. By etching metal samples with acid, Sorby revealed internal microstructures and correlated them with the properties and performance of the materials.

Transmission electron microscopy (TEM), a modern analytical cohort of metallography that reveals finer structural levels (see figure), suggests that the ancient swordmakers were accidental nanotechnologists. TEM analysis reveals nanometer-scale patterns of carbon in the hard edge of the sword. These patterns emerged from a process called "spinodal decomposition," in which a solid solution, such as the sword's high-carbon martensitic steel, becomes unstable and its constituents reorganize.[3] In modern parlance, the swordmakers' products could be described as "self-assembled heterophase nanostructures."

Inside steel. TEM reveals nanoscale structure of a quench-hardened steel. *G. B. Olson.*

The spatial dimension of materials is only part of what makes them tick. There is a spectrum of characteristic relaxation times associated with the various chemical and physical processes operating at the material's differing structural length scales. This adds the dimension of time.

The resulting dynamic spatiotemporal hierarchy means that any material at any time has structural features, such as grain sizes and dispersed particles, that have not yet reached equilibrium. That is why a material's structure and properties depend on how it was made and what conditions it endured in service.

Also fundamental to this dynamic view of materials is a realization that structural defects play vital roles at all length scales. Defects can make things fail in a thousand ways, but they're also often what makes materials so valuable. Perfect silicon crystal without dopant ions is not the semiconductor that has changed society. Pure iron metal without the right spicing of carbon

1906
Alfred Wilm discovers age hardening in aluminum alloy, which is later used for making dirigibles and other aircraft.

1909
Leo Baekeland patents Bakelite, the first entirely synthetic plastic, and commercializes it widely.

1911
Heike Kamerlingh Onnes discovers superconductivity in mercury chilled to temperatures near absolute zero.

1911–12
The father-son team of William Henry and William Lawrence Bragg, along with Max von Laue, develops the basis of X-ray crystallography, one of the most important analytic techniques for studying material structure.

1921
A. A. Griffith postulates role of defects in fracture strength.

Late 1920s
Hermann Staudinger argues that polymers are made of small molecules that link to form chains.

1931
Ernst Ruska builds first electron microscope.

1934
Wallace Hume Carothers invents nylon.

1940s
The wartime practice of organizing multidisciplinary research collaborations to achieve technological goals becomes a model for the subsequent organization of a field that later becomes known as materials science and engineering.

1947
John Bardeen, William Shockley, and Walter Brattain invent the transistor.

would never have become the steel backbone of the industrial revolution. Although "defect tolerance" remains a central tenet of modern materials science and is of incalculable commercial and safety importance, "defect engineering" is ascendant in the minds of many materials researchers. That's because defects on various hierarchical levels are a principal opportunity for controlling material behavior.

Because the personality of each material depends on all of these interacting spatial and dynamic attributes, it makes sense to approach materials as complex systems. Smith advocated such a systems view of materials decades ago. As more contemporary practitioners live by that insight, they are finding pathways to important new materials that can catalyze advances in manufacturing tools, computers, communications systems, and the myriad technologies whose very existence or improvement depends on more capable materials.

The Materials Discipline Comes of Age

Two principal branches of natural philosophy have evolved to form modern materials science. One is reductionist analysis, which takes nature apart to discern and understand its fundamental units. Reductionism has operated throughout the development of science. More sporadic has been the evolution of the synthetic systems view, which is better suited for understanding the connections holding nature together. In a new balance of these two philosophies, the systems view integrates the fruits of our investment in reductionism while replacing conventional discovery-based R&D with a far more effective and efficient design-based approach.

Although there were seeds of this turning point in materials science in early industrial laboratories at General Electric, Bell Labs, and elsewhere, the multifaceted field of materials research was deliberately synthesized as a single academic discipline in the late 1950s and 1960s with the founding of the first university materials departments.[4] Specialists in the science and technology of metals, ceramics, polymers, and composites collaborated in pursuit of unifying principles for the creation of materials of all classes. This meeting of minds helped lay the sociological and cognitive groundwork for a systems approach to materials.

The development of operations research in World War II and large-scale national missions such as the Manhattan Project and the Apollo space program were also important. It is estimated that at least 70 percent of our unprecedented economic boom of the past decade derives from technology, which in turn derives in good part from radically improved productivity via new systems-based methods of product development.

After World War II, newly formed government agencies, including the Office of Naval Research, the Defense Advanced Research Projects Agency, the National Science Foundation (NSF), and others, also helped midwife the multidisciplinary sensibility needed for developing complex materials. The same was true in academe. Starting with Smith's Institute for the Study of Metals at the University of Chicago, a series of university materials research laboratories established a national infrastructure for the interdisciplinary enhancement of materials science. The private sector added sociological foundations for the emerging discipline through a diverse mix of professional materials societies, as well as the establishment of numerous materials-centered journals and conferences.

The field now consists of many thousands of practitioners who share a growing sense of community, yet whose collective expertise runs a stunning gamut of materials categories. There are synthetic diamond makers, metal alloy designers, polymer scientists, optical fiber experts, thin film ceramic makers, developers of compound semiconductors, and "biomimetic" materials researchers who aim to emulate or adapt biology's unmatched brilliance in materials innovation. The list goes on and on, with a diversity akin to the living kingdom's millions of species.

Over the past four decades in materials R&D, however, there has been a consistent emphasis on "good science," as defined by reductionism, rather than "good materials," which emerge when engineering, manufacturing, and economic factors are included in the mix. This bias has limited the technological impact of the materials science community, particularly the academic portion. And materials engineering per se has been left primarily to industry, which has yet to fully benefit from the expanded base of materials science.

There is a twist in store. The science of materials has reached a level at which it now can radically change engineering practice. The possibilities are akin to what has come from the relationship between the life sciences and medicine. Until the nineteenth century, there was little or no science to guide medical technology and practice. Since then, however, the ever-growing corpus of biomedical knowledge has been leading to an ever more amazing stream of health care innovations. Here, a genuine desire to meet societal needs has produced a healthy mix of reductionist and systems viewpoints, yielding a culture that naturally integrates scientific understanding into practical use. The materials research community is poised to emulate this model.

1950s to 1960s

Much of the theoretical foundation behind the formation and evolution of material microstructure is developed. Among them is the Hall-Petch relation for grain refinement strengthening and the theory of diffusion of solids.

1953

Karl Ziegler develops catalysts that make it easier and cheaper to polymerize ethylene into stronger, more capable polymers.

1955

A team of scientists at General Electric combine high temperatures and enormous pressures to create synthetic diamonds.

1957

John Bardeen, Leon Cooper, and John Schrieffer provide theoretical basis for superconductivity, discovered in 1911.

1959
The U.S. government funds the first IDLs, or interdisciplinary laboratories, which mark a beginning of the modern academic model of materials science and engineering.
J. W. Cahn and J. F. Hilliard develop theory of microstructural evolution in diffuse-interface systems.

1970
Researchers at Corning develop optical fibers transparent enough to make fiber optic communication practical.

1974
A study by an NAS committee, COSMAT, defines field of materials science and engineering, creating a community sensibility.

1980
Gerd Binnig and Heinrich Rohrer invent scanning tunneling microscopy, which has led to a family of imaging tools often capable of molecular- and atomic-scale resolution.

1985
First university "materials by design" initiatives attempt computational materials design.

Materials by Design: Efficient Innovation

There is a general engineering design movement under way. It draws on the vast information pool generated by reductionist analysis, but adds the component of design, for which the systems approach is crucial.

Central to the materials design approach is a powerful logical structure connecting the "four elements" of materials science: processing, structure, properties, and technological performance. By connecting adjacent pairs of these elements, a three-link chain representing a versatile materials paradigm emerges. The deductive, cause-effect logic of reductionist science flows from processing to performance. All along the way, science reveals the relevant structures and phenomena, often in astounding clarity and detail. The inductive logic of systems engineering flows the other way, from performance to processing, thereby enabling designers to arrive at specific procedures likely to yield materials with the desired sets of properties and performance.

Not often successful in terms of producing useful new materials, early efforts at materials design nonetheless were harbingers of what materials development is to become. Generally, what has made or broken past efforts was whether they included the element of design. One notable success is the work of H.K.D.H. Bhadeshia and co-workers at Cambridge University, whose ambitious assault on the complex problem of weld metal design has spawned productive efforts at several national laboratories in the United States and Japan.

Materials by Design

To design a material is to try to meet a material user's need. A good place to start is with property cross-plots, like those of Michael Ashby, that graphically define property-performance relations.[13] These help engineers select materials for their product designs. They're useful for defining a quantitative set of property objectives that will sum into the materials performance needed by users. These performance specifications are determined by the role of the potential material in the wider system it serves. They also help define economic parameters, such as the cost of raw materials and processing, for the overall material design task.

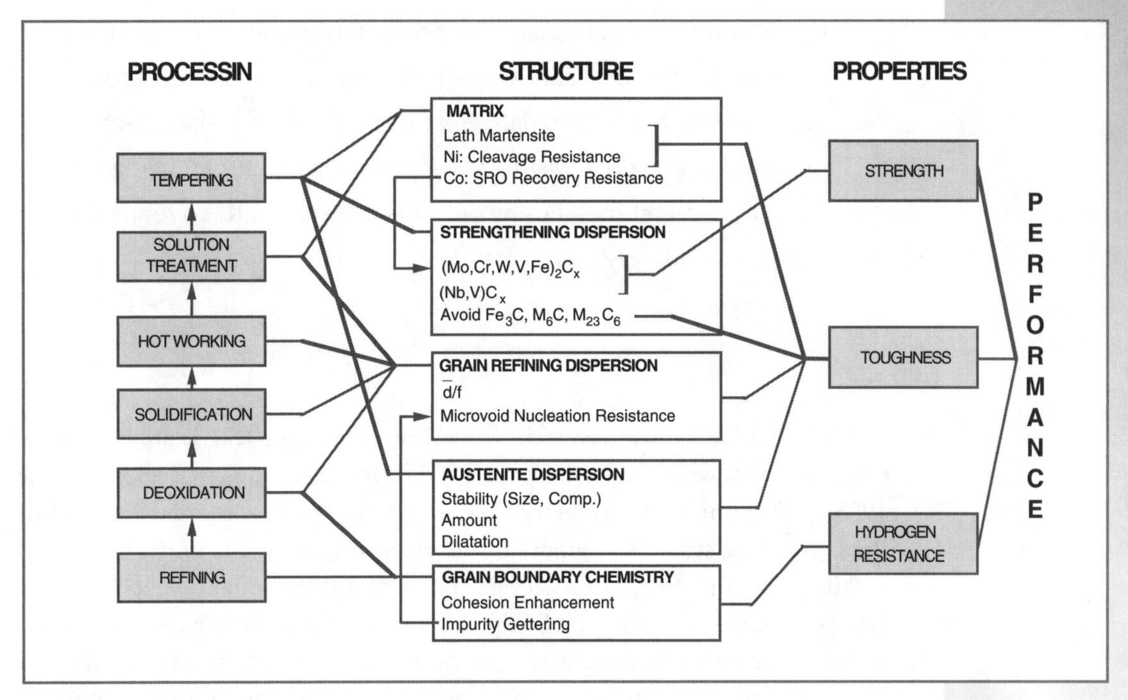

Planning materials. Flow-block diagrams guide materials design. *G. B. Olson.*

With objectives and economic constraints defined, the linear three-link framework of materials science and engineering (see figure) serves to guide the design and development phases. The Steel Research Group (SRG) (see main text) uses flow-block diagrams that represent 1) key microstructural subsystems (such as crystal grain sizes), 2) the primary links of these subsystems to the properties they control (such as strength and toughness), and 3) the stages of processing (such as tempering or reheating) that govern their dynamic evolution.

With these in hand, systems analysis is then applied to identify and prioritize key structure-property and processing-structure relations. Often, part of this exercise involves some additional modeling or empirical data gathering to fill gaps in the knowledge required for making practical decisions about composition and processing details. The systems view operates here at the strategic level, but it

1986

K. Alex Müller and J. Georg Bednorz discover high-temperature superconductivity in ceramic materials.

1990s

The field of materials science and engineering begins shifting into a more systems-based approach to materials innovation and toward materials design in which researchers can predict new materials they would like to have rather than having to discover them.

is supported by traditional reductionist analysis at a tactical level. On balance, the procedure greatly reduces the amount of costly experimentation in materials creation. Instead of making tens of prototypes along the way to a useful new material, SRG designers reach their target metal using only a few actual melts to refine the computation-heavy design efforts.

—G.B.O.

The initiative I know best is the Steel Research Group (SRG) at Northwestern University, which my colleagues and I have continuously developed since 1985.[5,6] This university-industry-government program was organized within the context of systems engineering to explore general methods, tools, and databases for the design of materials, using high-performance steels as a test case (see sidebar Materials by Design). There is a faith underlying this framework: The scientific knowledge base is now robust enough to supplant the traditional, empirically driven development of materials with a more efficient theory-driven and computationally based approach (see sidebar Virtualizing Materials).

In the SRG, we begin by combining the perspectives of materials users, suppliers, modelers, and designers into a set of specific materials property objectives. Those specifications, in turn, help us define how to use, adapt, or expand science-based models and databases of material behavior. We then use these models and databases to zero in on compositions and processing protocols that can transform those compositions into alloys fitting our specifications. In the past decade, we have used this framework to develop new alloys with unprecedented sets of properties. Some are now under evaluation by industrial and government partners for use in airplanes, power generators, aircraft carriers, and other applications.

Our own projects for designing steel are just a beginning. A recent NSF-sponsored workshop on Materials Design Science and Engineering[7] has called for broadening this approach to the design of all classes of materials. And in recognizing materials as one of five critical technologies for U.S. competitiveness, the President's Office of Science and Technology Policy[8] has identified computational materials design as a principal opportunity.

Opportunities abound for application of the new systems-driven computational design approach. Successful examples from our efforts include a stainless steel bearing for a space shuttle application, high-strength, high-toughness steels for aircraft landing gear and armor applications, and a new class of ultrahard steels for advanced gear and bearing applications. The more general validity of the design approach has been explored with computational design projects focusing on nonferrous alloys, hydrate ceramics, case-hardening polymers, and nanostructured thin film materials for microelectromechanical devices and hard coatings.

The new design capabilities will also help realize the dream of biomimetic materials, which emulate the complex adaptive microstructures of the living world that are beyond the reach of traditional empirical development. One successful demonstration is self-healing metallic composites. These incorporate "shape memory" alloys that exploit martensitic structural transformations to change their shape controllably. Integrating such components into thin film electronic devices to create efficient microactuators is leading to smart materials systems that unite the worlds of structural and electronic materials. Here the growing philosophy of predictive materials design is now combining the electrical engineer's realm of perfection-driven, artificially structured microcircuits with the materials traditions of self-assembly[9] and defect tolerance.[10]

Following the same philosophy can reduce the cost of discovery. In contrast to computational materials design, the prevalent industrial methods of materials development are based on an intrinsically slow and expensive process of trial-and-error empiricism. Theoretical input has been qualitative at best. It typically takes tens of millions of dollars over two decades to fully develop and qualify a new material in a critical application.[11] This sluggishness stands out in an era when engineers are expected to deliver new generations of products, such as automobiles, on an 18-month cycle. What's more, under these competitive pressures many industrial materials-development groups have been severely reduced or closed over the past decade. A recent National Research Council study[11] has concluded that the greatest challenge to the materials field today is the short time and cost constraints of the full materials development cycle.

Another reason these industrial materials development efforts have been downsized is anchored in the old R&D model in which new materials are discovered rather than designed. A typical estimate is that $1 of discovery costs $10 of development. If eliminating that first dollar was the only advantage of computational materials design, its impact on the total discovery and development cost would be small. The real advantage of materials design is that good design in the first place requires much less development later. And rather than using materials design approaches simply to provide an initial prototype for subsequent empirical development, there's yet more to be gained by integrating predictive modeling throughout the full design and development process.

Consider two costly and time-consuming phases of the standard materials development cycle: process optimization and qualification testing. A major concern in process optimization is scale-up. Because processing phenomena such as heat transfer depend on product size, a prototype material investigated on a small scale is not likely to behave the same way when processed on a large scale. "Solidification design" can preempt this common showstopper. By using models that simulate how materials respond to processing conditions for designing a material at the ultimate process scale of interest, we can reduce the number of expensive large-scale experiments while shortening the development cycle.

Improving qualification testing is more challenging. To design with confidence, designers must know how properties vary within the materials they intend to use. Current practice requires much testing to define these variations statistically. But most structure-property theories provide mean values of the variables of interest; they do not tell a designer about location-to-location variations within materials.

Help is coming, however. Researchers are developing a probabilistic materials science whereby structural distributions are mapped into property distributions. These are the kinds of data designers can use to make better predictions about how different materials will affect their products. Models of this kind already exist for fracture properties and heterogeneous phase transformations, and more efforts are being planned for this vitally important area.

Virtualizing Materials to Create Real Ones

Design of the hierarchical structure in materials requires a hierarchy of models based on materials science, applied mechanics, and even quantum physics (see figure). Models to help design features on the coarsest structural level of solidification, such as the chemical banding visible in the patterned sword blades mentioned earlier, employ powerful thermodynamic codes such as THERMOCALC.[14] These models enable designers to simulate the 10-mm scale of structure.

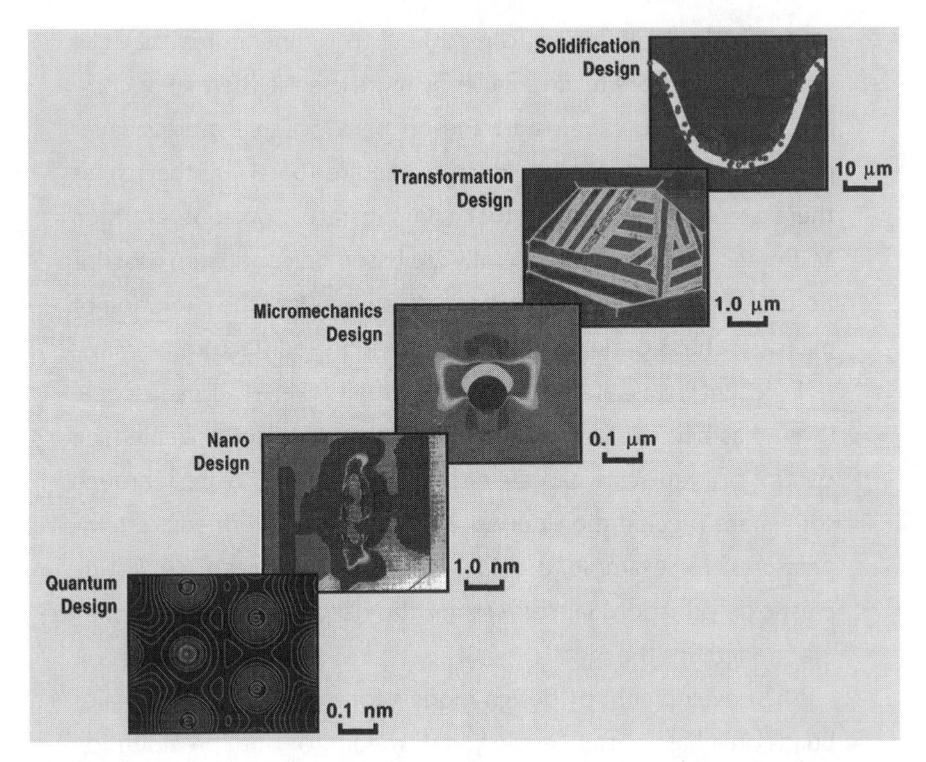

Model relay. Computational design of materials requires a hierarchy of models. *G. B. Olson.*

In metal alloys, this is the level at which the chemical partitioning between liquid and solid phases evolves during solidification processing. Application of these models aids decision-making about thermal processing details, a practice we like to call "solidification design."

At the 1.0-mm scale of structure, "transformation design" is the goal. This concerns the evolution of structural changes during quenching, whereby crystal grains present at high temperatures transform and subdivide as the hot alloy cools into a hierarchy of lower-temperature crystalline units. The design goal here is to specify and control processing temperatures so that desirable microstructures will form, while hindering the formation of competing microstructures that are less beneficial.

The 0.1-mm scale represents the micromechanics design level. An example of the phenomena relevant at this scale is "grain refining," in which the large grains formed at high temperatures that can embrittle alloys are made smaller by more precise thermal or compositional control. There's a trade-off here, because more smaller particles can also catalyze ductile fracture (breaking, that is), as there are more interfaces that can separate from one another. Micromechanics models typically are based on continuum descriptions of mechanical phenomena that can simulate the evolution of microstructures during material deformation and fracture.

In recent years, an even finer structural level—the nanoscopic level—has become better understood and more controllable. The control of 1-nm-scale particle dispersions in alloys created through solid-state precipitation during "tempering" at intermediate temperatures, for example, provides efficient obstacles for resisting plastic deformation. Said differently, this nanometer-scale structuring strengthens the metal.

The development of design models for these diminutive scales builds on a half-century evolution of theory for both precipitation and strengthening in metals. A major materials science breakthrough of the 1950s was the identification of dislocations as the key defects that enable the sliding of crystal planes, which shows up as plastic deformation of many materials. While most structure-property relations, such as the Hall-Petch relation for grain refinement strengthening, are based on empirical correlations, a major triumph of the 1960s was the Orowan particle strengthening equation. Derived directly from dislocation theory, it relates strength to the inverse spacing of dislocation obstacles (such as nanoscale precipitates). Such developments in theory, which mathematically codify materials behavior, are cornerstones for the claim that materials can be designed largely in silico.

The new accuracy of theoretical predictions of structure at the nanometer scale is only possible because of recent advances in high-resolution instrumentation allowing precise calibration and validation of theory in this regime. This includes such techniques as

X-ray and neutron diffraction, various electron microscopies, and atom-probe microanalysis. The latter is represented by the three-dimensional atomic reconstruction of a 3-nm strengthening carbide particle in an ultrahigh-strength steel. Such new capabilities in structural and chemical analysis down to the atomic scale open a new era of quantitative materials nanotechnology.

The electronic level is the finest level relevant to real materials. This is the realm of quantum design. As acknowledged by the 1998 Nobel Prize in chemistry shared by John Pople and Walter Kohn, the development of computational quantum mechanics and its extension via density functional theory constitute a profound advance that already has had significant industrial impact. A collaboration of materials science, applied mechanics, and quantum physics has enabled some of us in the Steel Research Group (see main text) to apply computational quantum mechanics to engineer steel at the subatomic level.

Our approach was to recast models of the impurity-induced embrittlement of grain boundaries into thermodynamic terms[15] and to rely on precise models of the atomic structures at grain boundaries. As represented by the computed valence charge density contours for a phosphorus atom at the core of an iron grain boundary, total energy calculations are sufficiently precise and accurate to explain the known effects of interstitial components such as boron, carbon, phosphorus, sulfur, and hydrogen on the cohesion of iron grain boundaries. What's more, the calculations lead to new mechanistic insights at the electronic bonding level. Extension to elements that occupy substitutional Fe sites in the boundary has enabled us to predict new alloying elements for enhancing boundary cohesion in steels. This is creating a new generation of "quantum steels" that incorporate these properties derived directly from electronic-level predictions.

—G.B.O.

Structure of Education

Education offers the most leverage for moving materials R&D into the systems-based paradigm. In its post-Cartesian form, modern metallurgy began with an emphasis on the direct correlation of processing and properties. The advent of physical metallurgy a century ago opened the "black box" of structure and brought a revolution in the fundamental understanding of the mechanistic link between processing and properties. This understanding created the foundation for the more recent generalization to materials science, making possible the general materials design and systems engineering methodologies described in this essay. Given the demonstrated potential of materials design, institutionalizing it via education would result in many benefits to society.

Due to the historic dominance of reductionist philosophy, however, we now have in place an analysis-oriented education system. In the name of objectivity, we train students to shut down the more synthetic and emotional tools of thought, which are precisely the ones best suited for doing good systems engineering. The future of engineering education, therefore, ought to target these subjective reasoning powers.

An activity of our new materials engineering curriculum at Northwestern, the Dragonslayer Project,[12] is designed to do this. The project explores the design of an "aesthetic" material through collaboration of freshman and upper-class design teams. The widespread presence of Western dragon combat in literature is used to integrate frontier steel technology with the history and legend of swordmaking to design an ultra-high-performance sword that our market analysis shows would be of maximum value to sword collectors.

Much of the mystique of the legendary Samurai sword, which even five centuries ago achieved performance levels equivalent to those of a modern carburized blade, stemmed from its ability to cut through other swords of the day. The equivalent performance advance has been adopted as a benchmark for property requirements.

Fanciful as the project might seem, it's no different from the kind of materials design that goes on at the DuPonts and Cornings of the world. Searches of literature, albeit medieval literature, were used to determine technical specifications, including the need for flame resistance when fighting fire-breathing dragons. Historians of the Dark Ages helped to identify and select a historically accurate sword style—the patterned double-edge broadsword. And market analysis helped determine what the target buyers, sword collectors, would be willing to buy and how much they might pay.

Finite-element mechanics simulations of cutting through a modern carburized blade led to a conceptual design for radical surface-hardening technologies to achieve this ambitious sword-cutting objective. We also were guided by the unlikelihood that anyone would face supernatural evil armed only with technology. This led to an additional design parameter, which market surveys indicated would constitute an important attribute in a collectible dragonslaying sword: The sword should be made from material of heavenly origin, namely, meteoritic iron. This is a technical feature with historical and legendary precedent (including the Excalibur legend).

Our systems approach to materials design has established the feasibility of using tabletop-scale aqueous electrolytic refinement of available meteoritic iron to achieve the necessary purity for producing the unprecedented alloy steel performance needed for a sword fit to defeat the most evil of adversaries. The freshman design team's proposed market plan includes auctioning a single "technomystical" sword for publicity, followed by commercial marketing of a range of high-performance steel products, including the "Dragonslayer golf club."

In this millennium, a new architecture of synthetic thought will continue its symbiosis with modern computational capabilities. And an age of empirical exploration will continue to be superceded by an Age of Design. This will open up powerful pathways by which human creativity, fused with scientific knowledge, will bring new levels of control over the material world widely applicable both to society's problems and its ambitions. The manifestation of the design paradigm, whether in an undergraduate or corporate setting, corresponds to a form of transmutation beyond the alchemist's dreams: the creation of materials from thought.

References

1. C. S. Smith, *A Search for Structure* (MIT Press, Cambridge, MA, 1981).
2. C. S. Smith, *A History of Metallography* (MIT Press, Cambridge, MA, 1988).
3. K. A. Taylor et al., *Metall. Trans.* 20A, 2717 (1989).
4. I. Amato, *Stuff: The Materials the World Is Made Of* (Basic Books, New York, 1997).
5. G. B. Olson, M. Azrin, E. S. Wright, Eds., *Innovations in Ultrahigh Strength Steel Technology,* Proceedings of the 34th Sagamore Army Materials Research Conference, 1990.
6. G. B. Olson, *Science* 277, 1237 (1997).
7. D. L. McDowell and T. L. Story, "New Directions in Materials Design Science and Engineering," National Science Foundation

Workshop Report, Georgia Institute of Technology Materials Council, Atlanta, GA, 1999.

8. "New Forces at Work: Industry Views Critical Technologies," 4th Office of Science and Technology Policy Report on Critical Technologies, February 2000.

9. G. M. Whitesides, J. P. Mathias, C. T. Seto, *Science* 254, 1312 (1991).

10. S. Williams, *Technol. Rev.* 102, 92 (September/October 1999).

11. D. Stein, *MSE: Forging Stronger Links to Users* (National Research Council, Washington, DC, 1999).

12. E. Davis, *Wired,* February 2001, pp. 136–43

13. M. F. Ashby, *Materials Selection in Mechanical Design* (Pergamon, Tarrytown, NY, 1992).

14. L. Kaufman and H. Bernstein, *Computer Calculation of Phase Diagrams* (Academic Press, New York, 1970).

15. J. R. Rice and J.-S. Wang, *Mater. Sci. Eng.* A107, 23 (1989).

6

Cloning: Pathways to a Pluripotent Future

ANNE MCLAREN

A gene is more of a big chemical than a living thing. Even an entire genome by itself isn't alive. It is at the cellular level that life begins . . . and assembles, and differentiates into the many tissues and forms that have yielded the Living Kingdom. In this essay, Anne McLaren follows the winding and sometimes pitted pathways that connect the origin of the cell theory of life in the 1830s to the new and unfolding era of cloning science and technology, as well as to the exciting biomedical arena of stem cells, which can be made to differentiate into different tissues. The possibilities are as fantastic as they are ethically and politically charged.

ANNE MCLAREN, a Principal Research Associate at the Wellcome/CRC Institute of Cancer and Developmental Biology in Cambridge, England has worked for many years on the genetics, reproductive biology, and developmental biology of mice. She undertook such work first for the U.K. Agricultural Research Council and then for the Medical Research Council. She is a member of the European Group on Ethics and advises the European Commission on social and ethical implications of new technologies. In addition, she's a member of the Human Fertilisation and Embryology Authority, which regulates in vitro fertilization and human embryo research in the U.K.

Cloning Timeline

1839
Theodor Schwann lays the foundation of what becomes known as the cell theory of biology.

1855
Rudolf Virchow states that all cells arise from cells.

1865
Gregor Mendel first reports the results of his pea plant experiments, from which he discerned his fundamental laws of heredity.

1885
August Weismann incorrectly proposes that the genetic information of cells diminishes as the cells differentiate during development.

1888
Wilhelm Roux helps initiate the field of experimental embryology by damaging cells of early embryos and observing the developmental consequences. The half-embryos that result seem to confirm Weismann's ideas.

Over the years, clones and cloning have meant different things to different people. Gardeners have always known that their vegetatively propagated plants formed a clone. There was a brief flurry of excitement in the 1970s when John Gurdon's spectacular photograph of 30 little albino frogs, cloned by nuclear transfer from an albino tadpole, reminded the newspapers of Aldous Huxley's *Brave New World.* The advent of recombinant DNA enabled molecular biologists to clone genes, but for most people, the word "clone" has had more to do with less expensive versions of IBM PCs than with anything biological. That wide usage of the term since the 1980s has helped change its meaning even in biological circles: "Clone" no longer signifies a group of identical members; it signifies a single member of such a group.

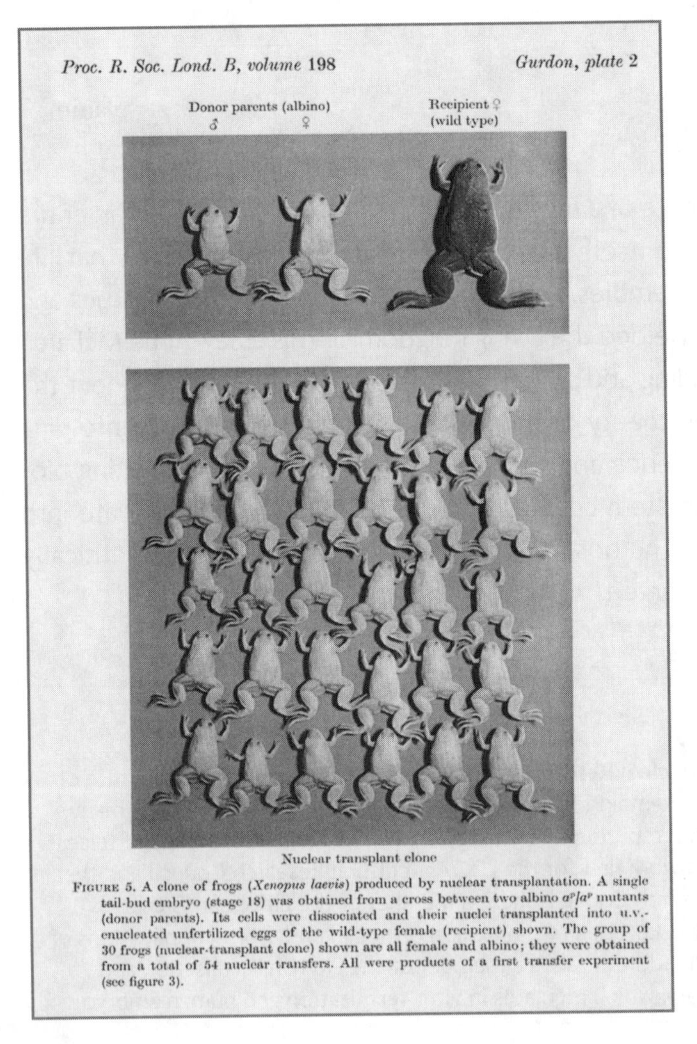

Proc. R. Soc. Lond. B, volume 198
Gurdon, plate 2
Donor parents (albino)
♂
♀
Recipient ♀
(wild type)
Nuclear transplant clone
FIGURE 5. A clone of frogs (*Xenopus laevis*) produced by nuclear transplantation. A single tail-bud embryo (stage 18) was obtained from a cross between two albino a^p/a^p mutants (donor parents). Its cells were dissociated and their nuclei transplanted into u.v.-enucleated unfertilized eggs of the wild-type female (recipient) shown. The group of 30 frogs (nuclear-transplant clone) shown are all female and albino; they were obtained from a total of 54 nuclear transfers. All were products of a first transfer experiment (see figure 3).

Frogs galore. This array of 30 cloned frogs spurred headlines in 1977. *J. Gurdon, Proc. R. Soc. London, Ser. B. Published by Royal Society of London.*

In 1997, cloning topped the charts of scientific and social discourse. That's when the news broke that Dolly, the Scottish cloned sheep, had been born. In reality Dolly represented just one stage in a whole series of experiments carried out in different labs by different teams of scientists, and all duly published in the scientific literature. But, for the general public, and indeed for many scientists whose attention was focused elsewhere, Dolly came like a bolt from the future. Because the nucleus that gave rise to Dolly came from an *adult* sheep (not even in frogs had an adult been cloned from adult cells), and because this feat of replication had been achieved in a *mammal,* the idea that *people* might also be cloned lost its air of fantasy. Making human clones became a real possibility.

There are two distinct scientific motivations that account for the creation of Dolly. The first is the fundamental desire to know whether the hereditary material in the nucleus of each cell remains intact throughout development, whatever the fate of the cell. The second relates in particular to farm animals: the ancient and ongoing desire to replicate those rare animals that possess an unusually favorable combination of genetic characteristics. The desire to augment those characteristics still further by genetic manipulation introduces still another interweaving strand—stem cell biology (see sidebar p. 121)—with its own history and its strong biomedical implications for the future.

In this chapter, I propose to consider Dolly not as a sheep but as a node, with scientific input streams flowing in, and scientific, social, and ethical consequences as outputs.

1892
Hans Driesch shows that each cell of a two-cell or four-cell sea urchin embryo can develop into separate, perfectly formed embryos, which goes against Weismann's theory. Roux's earlier results were likely due to damage from the hot needle.

1894
Jacques Loeb conducts early nuclear transfer experiment, in which the nucleus of one cell is transferred to a piece of egg cytoplasm containing no nucleus.

1914
Hans Spemann conducts nuclear transfer experiments with newts and later (1928) with salamanders.

The Role of the Nucleus

One of the questions that has inspired the science leading to and emerging from Dolly is: Does the hereditary material in the nucleus remain intact as the embryo develops? In other words, what role does the nucleus play in development?[1]

This part of the story really begins in 1839, when Theodor Schwann launched the cell theory, later to be encapsulated by Rudolf Virchow in his famous slogan, *Omnis cellula e cellula* ("All cells come from cells"). As

Biology cornerstone. Theodor Schwann laid the foundation of what became known as the cell theory of biology in 1839. *NLM.*

1932
Aldous Huxley publishes *Brave New World,* which describes a kind of human husbandry.

1938
In his book *Embryonic Development and Induction,* Spemann proposes a "fantastical" thought experiment: to introduce the nucleus from a differentiated cell into an egg whose own nucleus had been removed and then to see what would develop. The first of these experiments began 14 years later.

1952
Robert Briggs and Thomas King electrify the biological world by reporting the development of normal *Rana pipiens* tadpoles by transferring nuclei from embryonic cells to eggs from which nuclei had been removed.

1953
James Watson and Francis Crick report the structure of DNA.

applied to development, the cell theory requires two antagonistic properties: *cell heredity* and *cell differentiation.* Did every cell division produce two identical daughter cells, or did they differ? After the first cleavage division, could each cell on its own produce a whole embryo, or would one produce a left and one a right half, or one a front and one a back?

In 1888, Wilhelm Roux attempted to answer this question by damaging one cell of a two-cell frog embryo with a hot needle. The cell stayed in place but did not develop further; its partner developed into a left or right half-embryo. Sadly, this pioneering effort gave a misleading result, and August Weismann (who was more of a philosopher than an empirical scientist) used it to support his long-held, erroneous belief that all development and cell differentiation depended on loss of hereditary material.[2] Weismann's theory was soon disproved by Hans Driesch, who in 1892 separated the two-cell and even the four-cell sea urchin embryo into separate cells: Each developed into a small but perfect larva.[3] Similar results were obtained a few years later by others. One of them was Hans Spemann, who in 1901 found that if the first two cells of amphibian embryos were separated, two normal tadpoles could develop. It seemed that Roux's result was an unfortunate artifact due to the inhibitory effect of the damaged cell. Some invertebrates including nematodes, however, showed *mosaic* rather than *regulative* development: When separated, the first two cells really did have different fates.

The nucleus, containing the chromosomes, soon became recognized as the carrier of heredity. Was it also the engine of differentiation?

In order to explore nuclear, rather than cellular, potential, Spemann and Jacques Loeb carried out ingenious primitive nuclear transfer experiments in Amphibia and sea urchins, respectively. In both cases, the fertilized egg was constricted so as to give a portion containing the nucleus and a portion with-

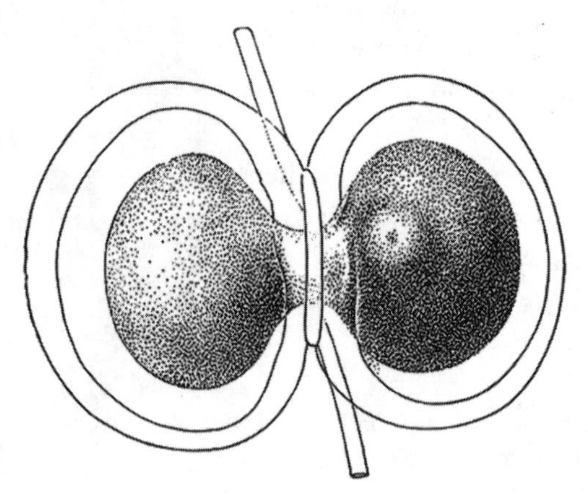

First steps. Using a hair to tie a fertilized frog egg into two halves, Hans Spemann took early steps toward cloning. *Spemann, Embryonic Development and Induction (Yale University Press, New Haven, CT, 1938).*

out. When the nucleated portion had cleaved to eight to 16 cells, one of these nuclei was allowed to reenter the portion of original uncleaved cytoplasm. Both portions were able to give rise to normal embryos, showing that the developmental potential of the nuclei remained unchanged at least to the 16-cell stage.

Experiments along these lines emboldened Spemann in 1938 to propose a "thought experiment," which, as he put it, "appears at first sight to be somewhat fantastical".[4] He wondered what would happen if a nucleus from a differentiated cell, even an adult cell, were to be somehow introduced into an egg whose own nucleus had been removed. It was 14 years before his *gedanken* experiment could be carried out. For one, Spemann lacked the know-how and techniques to carry out such a traumatic series of manipulations. And within a year, World War II had begun. It wasn't until 1952 that the necessary nuclear transfer technology was devised for Amphibia.[5] It took another 23 years before an adult amphibian nucleus was successfully transferred. And not until 1996, 58 years after Spemann articulated his fantastical experiment, was Dolly born.

Art Imitates Life

The image of thirty cloned frogs instantly conjures up thoughts of the Hatchery in Aldous Huxley's *Brave New World,* in which fertilized eggs destined to develop into Gamma, Delta, and Epsilon humans were subjected to "Bokanovsky's Process," each giving rise to up to 96 perfectly formed embryos, and each embryo developing into a fully grown adult.

Huxley's novel was published in 1932, six years before Hans Spemann's "fantastical experiment" and decades before John Gurdon's frogs. Where did Huxley's idea come from? No hint of cloning is seen in H. G. Wells's science fiction writings nor in Mary Shelley's *Frankenstein.* Huxley describes the Bokanovsky Process as a series of arrests to development, induced by X rays, followed by cooling, and finally alcohol. At each

1958
Michael Fischberg, Thomas Elsdale, and John Gurdon perform nuclear transfer in South African frogs leading to sexually mature adults.

1962
Gurdon announces cloned tadpoles using the nuclei of fully differentiated adult intestinal cells.

1963
J. B. S. Haldane uses the term "clone" in a speech titled "Biological Possibilities for the Human Species of the Next Ten-Thousand Years."

1970s
Fictionalized accounts of cloning technology proliferate. Examples include David Rorvik's *In His Image: The Cloning of a Man* (1978), Ira Levin's *The Boys From Brazil* (1976), about the cloning of Hitler, and Fay Weldon's *The Cloning of Joanna May* (1989), which illustrates how very unidentical human clones probably would be.

arrest the egg buds, then the buds bud and bud again like a plant, producing up to eight embryos each time. So perhaps it was plants that fired his imagination.

Other cloning novels appeared in the 1970s and 1980s. David Rorvik's *In His Image: The Cloning of a Man* (1978) purported to be a true story based on Derek Bromhall's nuclear transfer experiment in rabbits. (Bromhall sued Rorvik and the publishers.) Ira Levin's *The Boys from Brazil* (1976), about the cloning of Hitler, was made into a film. Fay Weldon's *The Cloning of Joanna May* (1989) is the best of these novels and illustrates how very unidentical human clones would probably be. Cloning literature is sure to expand in both the fiction and nonfiction genres.

—A.M.

1977
Gurdon's image of 30 cloned frogs sparks public coverage that associates cloning research with *Brave New World.*

1979
Steen Willadsen begins successfully rearing to adulthood single cells from eight-cell sheep and cattle embryos, in full awareness of the economic benefits that could accrue from rapid multiplication of genetically superior breeds.

1981
Karl Illmensee and Peter Hoppe claim to obtain live young via nuclear transfer using genetically marked inner-cell mass nuclei microsurgically

In 1952, Robert Briggs and Thomas King electrified the biological world by reporting successful nuclear transfer in the frog *Rana pipiens*[6]. The nuclei came initially from undifferentiated blastula cells; they were transferred to unfertilized eggs from which the nuclei had previously been removed. Once the eggs had been stimulated to develop, some produced normal tadpoles. Over the next few years, this group published a paper per year, each one detailing ever more ambitious experiments. When nuclei were taken from gastrula cells, the next developmental stage after blastulae, the proportion of normal tadpoles was much lower. From gut cells it was lower still, and nuclei from still later tail-bud stage embryos gave no normal development. By 1960, Briggs and King had concluded that differentiation was accompanied by progressive restriction of the capacities of nuclei to promote all the various types of differentiation required for normal development.

Meanwhile, across the Atlantic, the Swiss embryologist Michael Fischberg was working in Oxford with two younger colleagues, Thomas Elsdale and John Gurdon, on a different frog, *Xenopus laevis.* In many ways *Xenopus* was easier to work with than *Rana.* There was no need to remove the recipient egg nucleus, as it seldom took part in subsequent development. What's more, the researchers had a cell marker (the number of nucleoli) distinguishing donor and recipient strains, so there was never any doubt as to whether the nuclear transplantation had succeeded.

Already in their first paper, nuclei from early *Xenopus* blastulae were shown to support development not only to the tadpole stage, but through metamorphosis to give sexually mature adults.[7] Gurdon followed up this work: He found that nuclei from later stages could also support development to adults but less frequently—30 percent from blastulae, but only 6 percent from hatched tadpoles, and 3 percent from swimming tadpoles. Did these nuclear changes reflect what was going on in normal development, or were they the result of transplantation?

Experiments became more refined, and knowledge about what underlay the observed nuclear changes grew. Marie Di Berardino started looking at chromosomes. In 1967, she and King reported more than 1,200 transfers using *Rana* nuclei taken from differentiated neural cells. Only four out of the whole set had normal chromosomes, and three of these showed developmental abnormalities. They concluded that all the abnormal development they were seeing was attributable to chromosome aberrations occurring soon after transplantation. To give the nuclei a long period of exposure to the new cytoplasmic environment before they were required to replicate, Di Berardino started transplanting them to oocytes rather than to mature eggs. Eventually in 1983 Di Berardino and Nancy Hoffner showed that adult *Rana* red blood cell nuclei transferred to oocytes could support development up to the swimming tadpole stage. The same nuclei put into eggs got no further than the early gastrula.

In *Xenopus,* Gurdon greatly improved his success rate by doing serial nuclear transfers, rescuing normal nuclei from arrested embryos. He and his colleagues were able to produce fertile adult frogs using the nuclei of differentiated epithelial cells from the guts of feeding tadpoles. To prove that even nuclei from terminally differentiated cells had not lost their developmental potential, they showed that nuclei from specialized adult skin cells, identified with antikeratin antibody, could support development up to the swimming tadpole stage. That meant they must still retain the genetic information required for heart, muscle, brain, eyes, and all the rest.[8] These results were impressive, but still nobody had succeeded in making an adult amphibian by transplantation of an adult nucleus to an egg or oocyte.

Nuclear transfer experiments in mammals also had been going on. Live young were obtained from single blastomeres isolated at the two-cell stage in rabbits as early as 1952, and up to the eight-cell stage in rabbits in 1968. Following a few earlier attempts to induce development of enucleated mouse eggs by virus-induced fusion of somatic cells, Derek Bromhall obtained morulae by microsurgical introduction of early embryonic nuclei into enucleated

introduced into enucleated fertilized mouse eggs. But no one can replicate the experiments.

Several researchers report generating pluripotent embryonic stem cell lines from mouse blastocysts.

1983

James McGrath and Davor Solter succeed in obtaining young mice after transferring nuclei between unfertilized mouse eggs at the one-cell stage, using virus-induced fusion.

1984

McGrath and Solter fail to clone mice and claim that the cloning of mammals by simple nuclear transfer is impossible.

1986

Willadsen clones a sheep from embryo cells using nuclear transfer. This is the first such success to clearly stand the test of time.

1990

The Human Genome Project officially begins.

1995
Using differentiated cells of sheep embryos, Ian Wilmut and Keith Campbell of the Roslin Institute create Megan and Morag, the world's first sheep cloned from differentiated cells.

1997
Wilmut and colleagues report Dolly, the world's first creature to be cloned from adult cells.

President Clinton proposes a five-year moratorium on both federally and privately funded human cloning research.

Polly is born at the Roslin Institute. She is cloned from a fetal fibroblast into which had been inserted a gene for a valuable pharmaceutical protein, the human blood-clotting factor IX.

Richard Seed announces plans to clone a human being before regulatory laws could be enacted.

A wealthy Californian seeking to clone his dog Missy funds the "Missyplicity Project."

rabbit eggs.[9] Jacek Modlinski injected genetically marked nuclei into fertilized mouse eggs; development continued up to the blastocyst stage when nuclei were taken from morulae or from the blastocyst inner cell mass, but failed when nuclei from the already differentiated trophectoderm were used. Because the recipient eggs were not enucleated, the blastocysts were tetraploid and did not develop further.

The first claim to obtain live young after nuclear transfer in mammals was by Karl Illmensee and Peter Hoppe in the mouse using genetically marked inner-cell mass nuclei microsurgically introduced into enucleated fertilized mouse eggs.[10] The results have never been successfully repeated, despite determined efforts by James McGrath and Davor Solter. These two researchers succeeded in obtaining young after transferring nuclei between fertilized mouse eggs at the one-cell stage, using virus-induced fusion,[11] but nuclei taken later, even at the two-cell stage, were unable to support development.

Cloning for Replication

Embryo splitting in sea urchins, Amphibia, or mammals can give clones of two, four, or eight individuals. Neither Driesch nor Spemann carried out their experiments to increase numbers. Others did, however. From 1979 onward,[12] Steen Willadsen successfully reared to adulthood single cells from eight-cell sheep and cattle embryos, in full awareness of the economic benefits that could accrue from rapid multiplication of genetically superior breeds.

Nuclear transfer was not seen by Briggs and King as a means of replicating frogs. On the other hand, the first *Xenopus* nuclear transfer paper mentioned that a number of monozygotic frogs had been obtained from single donors, and in 1962 Gurdon published a picture of 20 cloned male frogs, of uniform color pattern. Two were small and sterile, the rest were of uniform size. By 1977, an albino *Xenopus* mutant became available and was used for the well-known picture of 30 small albino frogs.[8] They were made by transferring nuclei from a single albino tadpole into the eggs of a dark female.

Nuclear transfer experiments proved more successful in sheep[13] and cattle than they had been in mice. From the first, these experiments were designed to multiply the numbers of valuable animals rather than to examine the role of the nucleus per se.

In 1991 Willadsen and colleagues reported 101 nuclear transfer calves, using nuclei derived from cattle morulae. Further cattle studies yielded clones of up to eight calves ("octuplets") generated from a single donor embryo, and successes were reported with nuclei taken from cultured blastocyst cells. Unfortunately, many of the calves developed abnormally and were pathologically overweight at birth, so the procedure has not yet proved economical for cattle breeding.

1998
James Thomson and colleagues report deriving human pluripotential stem cell lines from supernumerary blastocysts donated by patients undergoing infertility treatment involving in vitro fertilization.

John Gearhart and colleagues report the derivation of human embryonic germ cell lines from aborted human fetal material.

The following Web site offers another relevant timeline: library.thinkquest.org/24355/data/timelinenav.html.

The Roslin Institute: Before and After Dolly

Ian Wilmut at the Roslin Institute in Scotland[14] was seeking a way to modify the genetic constitution of sheep and cattle more effectively than by the rather hit-or-miss method of injecting genes into the fertilized egg. Mouse embryonic stem (ES) cell lines made from the blastocyst inner cell mass were amenable to genetic modification, and nuclei from the inner cell mass of cattle had successfully been used to make nuclear transfer calves. Combining these two approaches offered a possible way forward. Keith Campbell, Wilmut's colleague, was impressed by the amphibian evidence that nuclei retained their full developmental potential even in differentiated cells. Also, he had worked on the cell cycle, and he was convinced that synchronization between donor and recipient cell cycles was the key to successful nuclear transfer.

The Roslin team first tried and failed to make immortalized and undifferentiated sheep ES cell lines. That frustration may have been an important factor in their subsequent successes. Unperturbed by the fact that the cells were differentiating, they continued to culture blastocyst inner cells. To optimize the chances of successful nuclear transfer, they put their cultured cells into a state of quiescence, which approximated the cell cycle stage of the recipient unfertilized sheep egg. Transfers were done, using electrical stimuli both to fuse the cultured cell with the enucleated egg and to kick-start embryonic development. From 244 nuclear transfers, 34 developed to a stage where they could be placed in the uteri of surrogate mothers. In the summer of 1995, five lambs were born, of which two—Megan and Morag—survived to become healthy fertile adults.[15] Megan and Morag were the first mammals cloned from differentiated cells.

The next season, Wilmut and Campbell became more ambitious. They repeated the transfers of nuclei from embryonic cells. They included a group using nuclei taken from cultured fetal fibroblasts, which give chromosomally stable cell cultures. And they also used nuclei from cultured mammary gland cells taken originally from a six-year-old ewe and stored, frozen, in liquid nitrogen. The first group produced four live lambs, the second two, and the third just one—Dolly.[16]

Megan and Morag. The first mammals cloned from differentiated cells. *Design & Print Services/The Roslin Institute.*

Dolly was the sole survivor from 277 transfers of adult nuclei. The procedure has now been extended to cattle, goats, pigs, and mice, but the success rate remains very low, seldom more than 3 percent. Of those that are placed in surrogate mothers, many die in utero. Others die at birth, often with abnormalities. The reason for this high mortality rate is not known. Perhaps the extensive reprogramming that the adult nucleus requires is incomplete.

By 1997 the project had moved on yet again. Polly was born, cloned from a fetal fibroblast into which had been inserted a gene for a valuable pharmaceutical protein, the human blood-clotting factor IX. Subsequent progress along these lines remains shrouded in commercial secrecy and the confused state of patent law.

Whither Cloning?

Just as the exploration of nuclear potential and the desire for replication have been two distinct strands in the evolution of cloning, so they remain distinguishable factors that will influence possible future lines of development, both in animals and in humans.

1. Replication.

In farm animals, cloning by nuclear transfer could replicate large numbers of genetically elite individuals that have highly advantageous combinations of genes, brought together either by traditional breeding methods, by transgenic technology, or by in vitro gene targeting, cell selection, and nuclear transfer. Without cloning, these unique gene combinations would be dissipated by genetic recombination. In the plant world, this approach is routine.

Cloning by nuclear transfer could be used for replicating household animals as well. Many people will likely request to clone their much-loved cats and dogs. A project to investigate nuclear transfer cloning in dogs, the "Missyplicity Project," has already been funded by a wealthy Californian seeking to clone his dog Missy. Pet owners may be disappointed, however. Genetic identity by no means ensures identity of personality or temperament.

When techniques of nuclear transfer cloning have been improved and extended, it might even be possible to recover species that have become extinct, provided some of their cells were preserved by freezing. This possibility provides a strong incentive to maintain tissue banks for endangered species.

Humans are animals, too, so what works with other animals will probably work with humans. Reproductive cloning, where a human embryo derived by somatic cell nuclear transfer is placed in a woman's uterus to develop into a baby, is out of the question at present. The large numbers of deaths and abnormalities that accompany reproductive cloning in mice and farm animals make any extension to humans totally unacceptable, at the moment anyway. But if the procedure became safer and more efficient, and if it worked in humans (it has not yet worked in rabbits), and if enucleated oocytes and surrogate mothers were available, what would be the ethical and social implications?

Many people find cloning human beings entirely unacceptable ethically, but there are many different reasons why people might wish to clone themselves or others. Some reasons seem more ethically objectionable than others.

The genetic constitution of anyone cloned from an existing person could not be confidential. Also, for the cloned child, the weight of expectation to be like his or her progenitor could become intolerably great. Neither of these objections would apply in the case of a couple who had finally achieved a pregnancy only to find that the fetus was ectopic, or the baby dead through an accident at birth. Nuclear transfer from the fetus or baby might offer their best chance of a replacement pregnancy. But any attempt to clone a talented violin player or a famous sportsman could lead to problems. The child might disappointingly dislike music, or despise sport, and be made deeply unhappy.

For couples whose infertility is so extreme that one partner is entirely lacking in germ cells, somatic nuclear transfer from one or the other might be their only means of having their "own" child. As the child grew up and came to resemble its progenitor (the nucleus donor) at the time when the couple first met, however, their relationship might suffer. For lesbian couples, or single women who have failed to find a man whom they wish to father their children, nuclear transfer from one of their own somatic cells, perhaps using their own eggs and their own uterus, would offer an interesting alternative to donor insemination or adoption.

Then there are people who fear death or desire immortality. From the ethical point of view, this would mean treating children as a commodity, merely as a means to an end, rather than desiring them for their own sake. But even with conventional reproduction, people have always had children for all sorts of reasons, not always for the sake of the children.

From the philosophical point of view, none of the ethical objections seem conclusive. The strongest arguments against human reproductive cloning are perhaps the social ones. It runs counter to our present culture, it would wreak havoc with family law, and, at least in Europe, public consultations have produced an overwhelmingly negative response: People don't want it.

2. The role of the nucleus.

We know almost nothing about how cloning by somatic cell nuclear transfer works. Increased understanding can come only from basic research using laboratory animals.

A differentiated nucleus has a set genetic program that has to be canceled and replaced by the genetic program of a fertilized egg at the very beginning of embryogenesis. How is this genetic reprogramming achieved? What are the crucial factors in egg cytoplasm? If telomere length[17] is not restored by reprogramming, does it matter? Is the mismatch between nuclear and mitochondrial genes ever a problem? Do some somatic nuclei reprogram more readily than others? Is the prior quiescence of the donor nucleus important? Why are there so many deaths and abnormalities, and so few born alive?

These questions will occupy scientists for many decades. If the cytoplasmic factors in the egg cytoplasm that are responsible for reprogramming are identified, it might be possible to reproduce them in vitro. Perhaps reprogramming could then be brought about without requiring the participation of a mature egg.

As basic research progresses, so too will technology. The genotype of cultured cells can be altered by recombinant DNA technology, including gene targeting to remove or replace genes. The use of nuclear transfer from such cultures to make animals of the desired genotype opens up many new perspectives in animal breeding. Not much is known about the genetics of quantitative characters such as growth and fertility, except that they are complex and probably under the control of many genes. Disease resistance could be an early target, but the most immediate impact of nuclear transfer cloning that we are likely to see will be animals producing valuable pharmaceutical products in their milk or even urine ("pharming"), or producing milk lacking the proteins to which babies are allergic, or milk or meat with enhanced nutritional value.

Stem Cells: Golden Opportunities with Ethical Baggage

If all cells come from cells, as Rudolf Virchow postulated in the 1850s, all but the most short-lived animals must harbor a reserve of cells to replace those that die or are damaged. This reserve consists of stem cells.[18] They are defined as those cells which can divide to produce a daughter like themselves (self-renewal) as well as a daughter that will give rise to specific differentiated cells. Stem cells in the body may be unipotent, like spermatogenic stem cells (which are responsible for the continuing production of spermatozoa), or they can be multipotent, like neural or hemopoietic stem cells, which give rise respectively to all the varied cell types in the nervous system or in the blood and immune system. Given the possibility of directed differentiation of stem cells, these multipotent somatic stem cell lines may prove to be of significant clinical value.[19]

Experimentally, it has also proved possible to create immortalized pluripotent stem cells. In 1981, pluripotent embryonic stem (ES) cell lines derived from mouse blastocysts were reported.[20] These will proliferate indefinitely in vitro as undifferentiated cells, but will also differentiate when the culture conditions are modified, and when introduced back into an embryo, they will successfully colonize every cell lineage including the germ line. However, pluripotent stem cells cannot on their own make an embryo, that is, they are not totipotent. Undifferentiated ES cell lines have been extensively used in mice for genetic manipulation, including the introduction of new genetic material as well as knocking out and replacing genes. Later, similar pluripotential stem cell lines were derived from mouse embryonic germ (EG) cells.[21] Despite energetic attempts, it proved extremely difficult to make ES or EG cell lines from any species other than the mouse.

That changed in 1998 when James Thomson and colleagues in Wisconsin reported that they had derived human pluripotential stem cell lines from surplus blastocysts donated by patients undergoing infertility treatment involving in vitro fertilization.[22] In the same year, John Gearhart and colleagues reported the derivation of human EG cell lines from aborted human fetal material.[23] All these lines are now owned by Geron Corp. of Menlo Park, California; some others have been made elsewhere and are being studied in Australia.

Intense activity is now being focused on both mouse and human pluripotential stem cells, in an attempt to induce directed differentiation to defined cell types in culture, for example, by exposing the cells to signaling molecules such as retinoic acid and cytokines, as well as by genetic manipulation.[24] The ultimate aim here is to supply transplant surgeons with a readily available supply of any tissue for the repair of damaged or diseased organs so that the need for organ donors would drop. Harold Varmus, until recently director of the National Institutes of Health (NIH), stated before Congress: "There is almost no realm of medicine that might not be touched by this innovation." Among the many medical possibilities are the use of cardiac muscle cells for heart problems, pancreatic islet cells for diabetes, liver cells for hepatitis, and neural cells for Parkinson's or Alzheimer's disease. In animal models, some successes have already been achieved: ES cell-derived cardiac muscle cells have been incorporated into damaged rat hearts, and neural cells introduced into the brain of a mouse model of multiple sclerosis have differentiated into appropriate cell types.[25]

In mice, EG cells introduced into embryos have led to some abnormalities, so they may be less suitable than ES cells for clinical use.[26] ES cells raise ethical problems, however, as they are derived from early human embryos. Some people believe that fertilized human eggs and early embryos are already persons. They will therefore object to their use for research, even for such ends as cell and tissue therapy to reduce human suffering and disease. Others argue that, because the donated blastocysts will never be transferred to a uterus, it is preferable for them to be used for a beneficent purpose than to merely be left to perish. NIH is now prepared to fund research on human pluripotential stem cells that have been derived according to certain guidelines, but they will not fund the derivation of such lines.

—A.M.

New medical treatments may be the most exciting single outcome of cloning by somatic cell nuclear transfer. If the biomedical uses of pluripotent human stem cells can be realized, cell and tissue therapy for many serious diseases will become available. But because the patients may still reject the transplanted cells, they will have to take immunosuppressive drugs, along with the associated risks of infection and cancer. Maybe the patients could be rendered tolerant, or maybe the cells could be genetically manipulated to make them nonantigenic.

Or maybe not. The alternative would be to use somatic cells from the patients themselves for nuclear transfer, so that the early embryo and any pluripotent stem cells derived from it would be genetically and antigenically identical to the patient, 100 percent compatible. No question of transplant rejection could then arise. This approach is certainly not on the immediate agenda and would require a fair amount of prior research, but it appears technically feasible and could greatly reduce suffering. Because no reproductive cloning is involved, the ethical objections outlined earlier would not apply.

There would of course still be people who believe that personhood is present from the very beginning of embryonic life, so that using an embryo for any purpose other than making a baby is tantamount to murder. The stroke victim, the multiple sclerosis patient, the person crippled with rheumatoid arthritis may, on the other hand, believe that they have every right to use what are effectively their own cells.

The twenty-first century will see many deep ethical conflicts, but it will also see unprecedented biomedical advances that will benefit all humankind.

References and Notes

1. For insight into the early history of embryology and genetics, two books written in the 1930s by outstanding scientists cannot be bettered: T. H. Morgan, *Embryology and Genetics* (Columbia University Press, New York, 1934), and J. Needham, *A History of Genetics* (Cambridge University Press, Cambridge, 1934).
2. A. Weismann, *The Germ-Plasm Theory of Heredity* (Charles Scribner's Sons, New York, 1902).
3. H. Driesch, *Anat. Anz.* 8, 348 (1893).
4. H. Spemann, *Embryonic Development and Induction* (Yale University Press, New Haven, CT, 1938).
5. For an on-the-spot account of one group's activities in this field, as well as a more general overview, M. Di Berardino's *Genomic Potential of Differentiated Cells* (Columbia University Press, New York, 1997) is recommended.
6. R. Briggs and T. J. King, *Proc. Natl. Acad. Sci. U.S.A.* 38, 455 (1952). For later developments, see R. Briggs and T. J. King, *Dev. Biol.* 2, 252 (1960).
7. J. B. Gurdon, T. R. Elsdale, M. Fischberg, *Nature* 182, 64 (1958).
8. J. B. Gurdon, *Proc. R. Soc. London, Ser. B* 198, 211 (1977).
9. D. Bromhall, *Nature* 258, 719 (1975).
10. K. Illmensee and P. C. Hoppe, *Cell* 23, 9 (1981).
11. J. McGrath and D. Solter, *Science* 220, 1300 (1983).
12. S. M. Willadsen, *Nature* 277, 298 (1979).
13. S. M. Willadsen, *Nature* 320, 63 (1986).
14. For a fuller account, see I. Wilmut, K. Campbell, C. Tudge, *The Second Creation: The Age of Biological Control by the Scientists Who Cloned Dolly* (Headline, London, 2000).
15. K. H. S. Campbell, J. McWhir, W. A. Ritchie, I. Wilmut, *Nature* 380, 64 (1996).
16. I. Wilmut, A. E. Schnieke, J. McWhir, A. J. Kind, K. H. S. Campbell, *Nature* 385, 810 (1997).
17. Telomeres are structural features at the ends of each chromosome. They shorten with successive cell generations until the chromosomes can no longer function normally. Telomere

length can be restored by an enzyme, telomerase, present in some germ line cells and some stem cells.

18. Special issue on Stem Cell Research and Ethics, *Science* 287, 1417–1446 (25 February 2000).
19. B. D. Yandava, L. L. Billinghurst, E. Y. Snyder, *Proc. Natl. Acad. Sci U.S.A.* 96, 7029 (1999).
20. M. J. Evans and M. H. Kaufman, *Nature* 292, 154 (1981); G. B. Martin, *Proc. Natl. Acad. Sci. U.S.A.* 78, 7634 (1981).
21. Y. Matsui, K. Zsebo, B. L. M. Hogan, *Cell* 70, 841 (1992); J. L. Resnick, L. S. Bixler, L. Cheng, P. J. Donovan, *Nature* 359, 550 (1992).
22. J. A. Thomson et al., *Science* 282, 1145 (1998).
23. M. J. Shamblott et al., *Proc. Natl. Acad. Sci. U.S.A.* 95, 13726 (1998).
24. M. Li, L. Perny, R. Lovell-Badge, A. G. Smith, *Curr. Biol.* 8, 971 (1998).
25. O. Brüstle et al., *Science* 285, 754 (1999).
26. T. Tada et al., *Dev. Genes Evol.* 207, 551 (1998).

7

The Quickening of Science Communication

ROBERT LUCKY

Communication with other researchers is the lifeblood of scientific progress. In this essay, Robert Lucky, a seasoned essayist as well as a technical and entrepreneurial participant in the current communications boom, examines how new communications technologies have had on the way science is done. He also reflects on the ways new scientific discoveries, such as laws of electromagnetism, have steered the evolution of communications technology. In so doing, he sets the stage for considering how the awesome convergence of the internet, computational power, and communications bandwidth may play out in the coming decades.

ROBERT LUCKY is Corporate Vice President of Applied Research at Telcordia Technologies, based in Red Bank, New Jersey. He joined Telcordia in 1992 after an extensive career at Bell Laboratories. He is an inventor and author in the field of communications and is well known for his bimonthly columns in *Spectrum,* which is published by the Institute of Electrical and Electronics Engineers (IEEE).

TIMELINE OF SCIENCE AND COMMUNICATIONS

3500–3100 B.C.
Sumerians develop cuneiform writing using a stylus to etch wedge-shaped symbols into soft clay. Meanwhile, Egyptians develop hieroglyphic writing and use papyrus, the physical and etymological precursor of paper.

In the sixteenth century, science progressed at the pace of the postal system. Often it would take six months for one scientist to learn of the ongoing results of another. It took even more time for scientists to build on one another's accomplishments. The great Danish astronomer Tycho Brahe meted out his results more carefully than anyone before him, yet it was only after Brahe's death that Johannes Kepler was able to inherit the observational data he used to discern his laws of planetary motion. Today astronomers post data on the Web for instant worldwide access, and they routinely manipulate telescopes remotely through the Internet. Communication has always been the circulatory system of science, if not its very heartbeat. Yet even as progress in communications technology speeds the progress of science, there is a recursive relationship in which science improves communications. There is both a communication of science and a science of communications.

Communication Before Electricity

Today communications technology is almost synonymous with the electrical means of communicating: radio and television, the telephone and Internet. Yet science itself—defined as the quest for knowledge about nature—far predates the electrical era of the last century. The practitioners of science already were at work during the early evolution of written language, which is the fundamental underpinning of modern human communication. After all, the builders of Stonehenge demonstrated a remarkable knowledge of engineering and astronomy around 3100 B.C. This is almost contemporary with the beginnings of the Sumerian written language and surely predates the first syllabic writings, in the languages Linear A and B. Usage of Linear A, the older of the two, dates to about 1800 B.C.

Hardware. In 3500–3100 B.C. Sumerians developed cuneiform writing using a stylus to etch wedge-shaped symbols into soft clay. *Araldo de Luca/Corbis.*

After the development of written language, the next stage of communications was the emergence of postal systems, which could transport that recorded language farther than the loudest voice. Historical references to postal systems date back to 2000 B.C. in ancient Egypt and to about 1000 B.C. in China. These ancient systems invariably employed relays of messengers, either mounted or on foot, and were a natural outgrowth of the need for administration of states and kingdoms. The most celebrated ancient postal system was the *cursus publicus*, which is credited with holding together the vast Roman Empire over a number of centuries before its disintegration.

Although a postal system provides the rudiments of individual communication, it does not enable broadcast or widespread knowledge dispersal. For that purpose, books have been a remarkable social instrument, and they've remained unchanged in concept from nearly the time written language arose until today. There are early books on clay tablets in Mesopotamia and Egyptian papyruses that date from the third millennium B.C. Without a system for access, however, the power to communicate via books was limited, which is why the rise of the library was so important. This, too, was an ancient innovation, dating back to at least the third century B.C. with the founding of the most famous library of antiquity, the Library of Alexandria.

Books had existed for 2000 years before the invention of movable type and the publication of the Gutenberg Bible in 1454. After the conception of written language, this may be the most important invention in the history of communications in general, and in science communication in particular. The printing press made possible the widespread dissemination of books, which

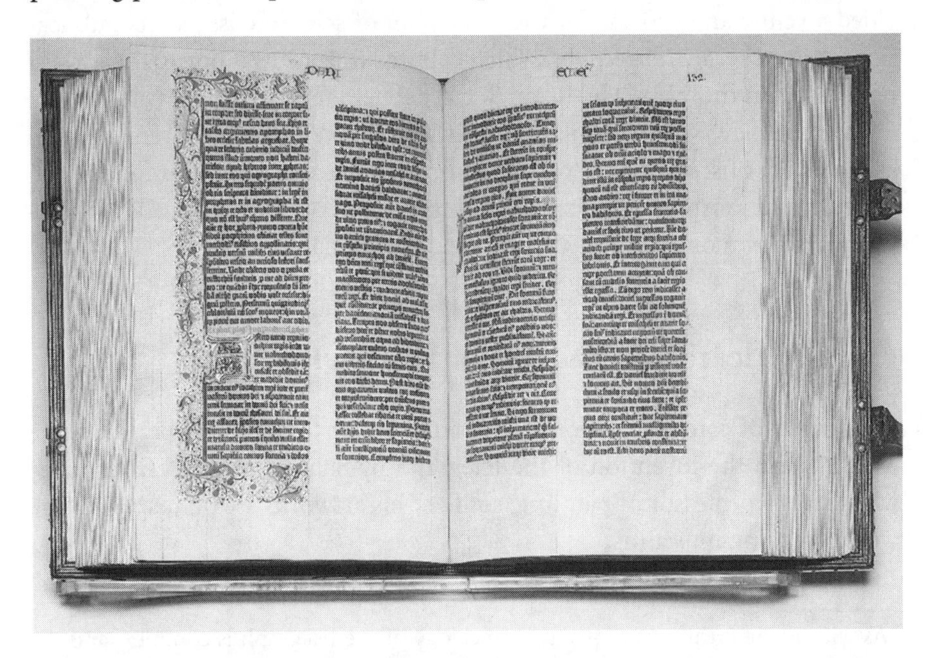

The Good Book. The printing of the Bible by Johann Gutenberg marked a new era in the dissemination of information. *The Pierpont Morgan Library/Art Resource.*

2000 B.C.
Early postal system is established in Egypt.

c.1800 B.C.
Minoans write using Linear A, the first known syllabic script.

300 B.C.
The founding of the Library of Alexandria helps establish practice of collecting knowledge and making it more widely accessible.

A.D. 1454
Johann Gutenberg uses movable type and a press to produce 300 bibles.

1500s–1600s
Postal systems proliferate in Europe.

1609
First regularly published newspaper appears in Germany.

1660
The Royal Society is founded and begins publishing its *Philosophical Transactions* seven years later. They become models for other scientific societies.

1700
Gottfried Liebniz shows that a binary numeric system of 1's and 0's can be used to denote any number.

1732
Benjamin Franklin starts a circulating library.

1832
Charles Babbage conceives of the "Analytical Engine," forerunner to the computer.

1837
Samuel Morse patents his version of the telegraph and a year later creates his now-famous code of long and short electrical pulses to represent letters. In 1844, a telegraph connects Washington and Baltimore, and the following message is sent:
"What hath God wrought?"

1847
George Boole mathematizes logical arguments.

1866
After a short-lived transatlantic cable was laid in 1858, the first reliable one is installed.

had previously been the exclusive property of the rich and powerful. The subsequent history of science is filled with books of singular fame and influence, such as Newton's *Principia Mathematica* and Darwin's *Origin of Species.* Even in the fast and transient world of today, science is still marked by the publication of a succession of celebrated books that summarize the wisdom emerging from the ongoing multitude of research papers.

Books and journals have been the carriers of information in science over the centuries, but there is, in addition, a social infrastructure necessary for effective communications in any community. It may seem remarkable in reading histories of science that the scientists working in a particular field always seem to know each other and to correspond regularly, in spite of the seemingly insuperable barriers between countries and continents in the ages before airplanes, telephones, and radios. The world of a particular scientific field has always seemed to be a small place. The rule of six degrees of separation has been especially effective in science, where fame and acceptance of one's views by peers have been powerful motivating forces. Even today the millions of scientists worldwide naturally subdivide themselves into specialties where everyone knows everyone else. Science has become a busy system of small worlds.

The social infrastructure for the communication of science was first formalized with the founding of the Royal Society in London in 1660. This provided a venue and process for the discussion of scientific issues. Its publication, *Philosophical Transactions,* which debuted in 1667, was one of the earliest periodicals in the West. The Royal Society's members, such as Isaac Newton, Edmond Halley, Robert Hooke, and Christopher Wren, are still remembered as the giants of scientific history. In contrast, science today is brimming with societies and journals in which most of the names on the mastheads are relatively unknown outside of their own fields of specialization.

Communication in the Age of Electricity

The science of communications is generally understood to have begun as late as 1837 with the invention of the telegraph by Samuel Morse.* (Curiously, Morse was a professor of painting; some of his artworks are on display in the world's leading museums.)

*As with many great inventions, the genealogy of the telegraph is complex, and the attribution to a single inventor at a particular moment in time is a great oversimplification.

The major significance of the telegraph was that it severed the bond between transportation and communication. Before the invention of the telegraph, information could move only as fast as the physical means of transportation would allow. It took weeks before what was known in Europe became known in America. The telegraph enabled a kind of space-time convergence that brought countries and continents together, causing profound effects on economics and government policies. At one stroke, the life force of science—information—was freed of its leaden feet and allowed to fly at the speed of light. Conceptually, at least, advances in communications since then have simply made instant communication easier and more convenient.

In fact, the telegraph was awkward to use. Morse's famous code, which is a relatively efficient representation of the English alphabet, was a practical help, but it required transcribing natural language into another symbol system. (More importantly, it presaged an outpouring of theoretical studies of coding in the field of information theory nearly a century later.) Still, the telegrapher's experience was not much different from that of the writer using the postal service: Both had to compose carefully crafted messages and take them to the nearest telegraph or post office. In contrast, Alexander Graham Bell's telephone brought interactive and easily accessible communications directly to the user. At the time of Bell's invention in 1875, the United States was already densely wired with telegraph lines, and the Western Union Co. was one of the largest corporations on Earth.

The telegraph, of course, was undone by the telephone. Bell's invention was basically the idea of analog—that is, the transmitted voltage should be proportional to the air pressure from speech. In that era it was the right thing to do. After all, the world we experience appears to be analog. Quantities such as time, distance, voltage, and sound pressure seem to be continuous in value, whereas bits—ones and zeros, or dots and dashes—are artificial and seemingly unrepresentative of reality.

1873
James Clerk Maxwell publishes theory of electromagnetism.

1876
Alexander Bell patents the telephone.

1888
Heinrich Hertz observes radio waves.

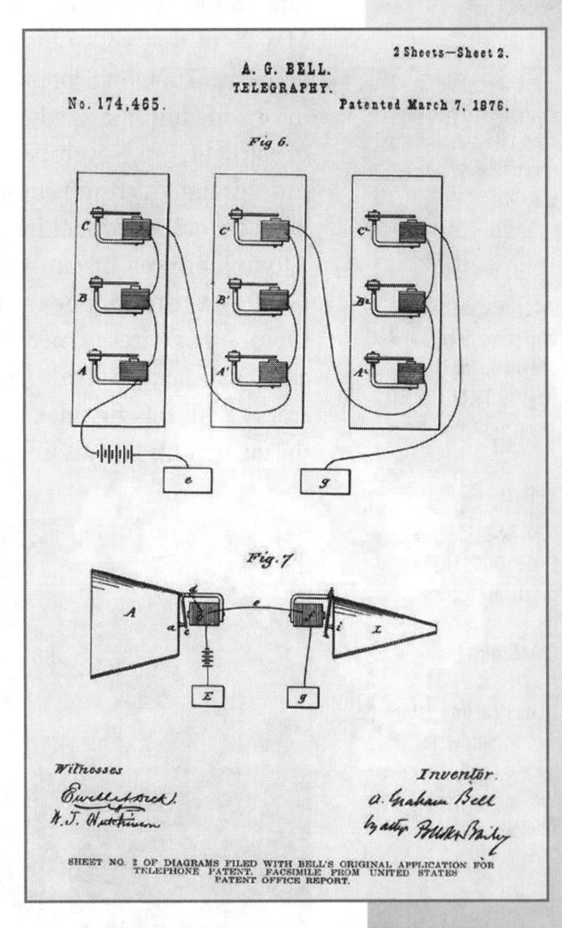

Bell's big patent. The ascent of telephony began with this 1876 document. *AT&T Archives.*

1894
Guglielmo Marconi invents wireless telegraphy—radio. Seven years later, he sends a radio signal across the Atlantic Ocean.

1924
With 2.5 million radio sets in the United States, radio enters its Golden Age.

1925
Commercial picture facsimile radio service is established across the U.S.

1928
Television sets are put into three homes and programming begins. Regular TV broadcasts begin in 1939.

1946
John Mauchly and John Eckert build the first practical electronic digital computer—ENIAC, which stands for Electronic Numerical Integrator and Computer. It weighs nearly thirty tons and takes up 140 square meters of space.

1947
William Shockley, John Bardeen, and Walter Brattain invent the transistor at Bell Labs.

For over a century thereafter, the wires that marched across the nation would be connected to telephones, and the transmission of voiced information would be analog. Progress in telephony had mainly to do with extending coverage across the nation and meeting the social goal of universal service. The functionality of the telephone remained almost unchanged until the latter part of the 20th century.

Whether the pace of science was immediately changed by the telegraph and telephone is difficult to determine. The handling of the British expeditions during the solar eclipse of 1919 to measure deflection of starlight at the sun's perimeter (a test of Albert Einstein's predictions based on his general theory of relativity) illustrates how the pace of science generally continued at a postal rate and yet could, on occasion, be electrified. Although the eclipse was on May 29 of that year, Einstein had still not heard of the results as late as September 2. Then on September 27 he received a telegram from Hendrik Lorentz informing him that Arthur Eddington had found the predicted displacement of starlight. On November 6 there was a famous meeting of the Royal Society proclaiming the results more widely, but this time, because of the telegraphed news of that meeting, Einstein awoke in Berlin on the morning of November 7 to find himself instantly famous in the newspapers of the world.

While the telephone companies were wiring the world, there was a parallel evolution of wireless media beginning with Guglielmo Marconi's experiments in 1894. Marconi made James Clerk Maxwell's theoretical work a practical reality with this first demonstration of radio transmission. In 1901 he defied the current understanding of line-of-sight transmission by successfully trans-

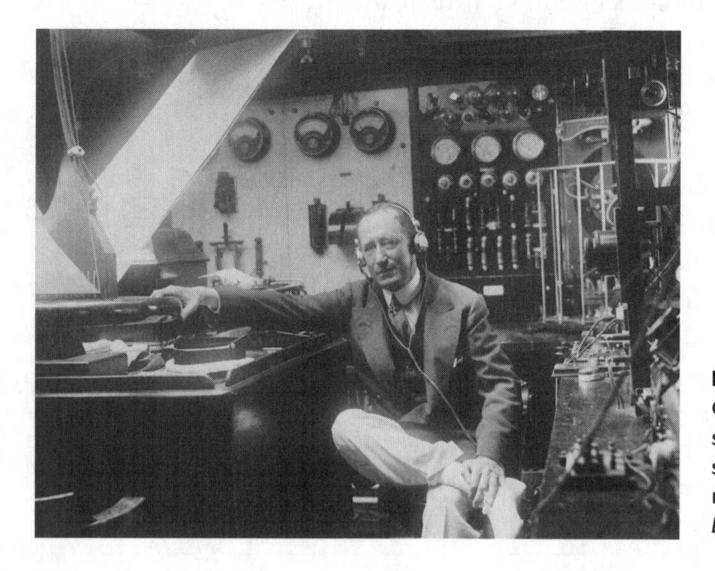

Father of radio. Guglielmo Marconi sits comfortably surrounded by his radio gear. *Bettmann/Corbis.*

mitting a signal across the Atlantic Ocean from St. Johns, Newfoundland, to Cornwall on the English coast. At that time there was no understanding of the importance of the reflective properties of the ionosphere in radio propagation. Viewed in retrospect, this was the kind of remarkable achievement sometimes made by determined amateurs who refuse to accept expert opinion.

In one of those curious juxtapositions of historical incidents, the sinking of the *Titanic* propelled radio into the limelight. Signals from the spark transmitter on the *Titanic*, picked up by neighboring ships, resulted in many lives being saved. Even more could have been saved had the nearest ship been attending to its radio, a fact that soon led to governmental regulations. A young telegrapher, David Sarnoff, came to fame by broadcasting the news of survivors; he later went on to head RCA and to play a major role in the world-changing developments of both radio and television.

Television was certainly the most influential communications medium of the last century. There came to be more television receivers in the world than telephones, and television broadcasts have purveyed culture to the farthest corners of Earth. The popularity and politics of science were affected by this new medium. The television broadcast of Neil Armstrong stepping on the moon on July 20, 1969, was a singular event in history, when people throughout the world were unified in their celebration of science. Moreover, educational television made heroes of individual scientists, such as Jacob Bronowski in *The Ascent of Man*, but especially Carl Sagan in *Cosmos*. Sagan became an icon of science, and with his death in 1996 the last popularly known scientist disappeared from public view.

As the wireless media expanded, so did the capacity of the terrestrial telecommunications facilities. What began as a single telephone conversation on a pair of copper wires has evolved into several hundred thousand conversations being carried by optical pulses over a glass fiber. Along the way, the copper wires were first supplanted by microwave radio relay systems. These relied on technology developed for radar during World War II, incorporating klystron and magnetron generators first developed in England. However, the microwave radio systems were soon surpassed in capacity by transmission over buried coaxial cables. Then, around 1960, Bell Labs began developing a millimeter waveguide system that was projected to meet all the demands for communications capacity well into the next millennium.

The millimeter waveguide system used a cylindrical metal pipe about 5 cm in diameter, with the transmission mode such that the electrical field was zero at the inside circumference of the guide. In theory, there was no loss with distance in such a mode, but that only held true as long as there were no bends

1949
Claude Shannon publishes landmark work on information theory, which becomes a theoretical cornerstone for the subsequent Information Age.

1958
Jack Kilby demonstrates the integrated circuit by fabricating several transistors onto a single substrate.

1960
Theodore Maiman builds the first practical laser, which had been conceived of earlier by Charles Townes, Arthur Schawlow, and others.

1962
The U.S. launches Telstar, the first true communications satellite. It could receive and then amplify and resend radio signals.

1965
Gordon Moore articulates what has become a famous law of technological development: There is a factor-of-two improvement in semiconductor technology every eighteen months.

1969
ARPAnet, the seed of the Internet, becomes operational.

1970
A team of Corning researchers makes practical optical fibers that are transparent enough for communications.

1971
Intel builds the microprocessor, "a computer on a chip."

1973
Robert Kahn and Vinton Cerf formulate the Internet-defining file-transfer protocol, TCP/IP.

1975
The first personal computer, the Altair 8800, is marketed.

1980
CNN begins 24-hour news channel.

1980s to present
As computational power increases, computational science ascends to become a third arm of science, joining theoretical science and experimental science.

or imperfections. In those days, the capacity of this system was considered enormous, and such capacity would be needed for what was considered to be its dominant future usage—the Picturephone, which was undergoing final development at Bell Labs during the same time period of the late 1960s. Neither of these major developments were ever a commercial success. The Picturephone, introduced commercially by AT&T in 1971, was a market failure. And the millimeter waveguide system was abandoned when it suddenly became evident that a fantastic, newly developed optical fiber offered higher capacity and lower cost.

Those optical fibers were key components of a leapfrog technology that had been in the wings for more than a decade. The laser had been invented by Arthur Schawlow and Charles Townes in 1958, and Theodore Maimen had built the first practical laser two years later. At first its use for communications had been envisioned as modulated beams in free space. At the time of the millimeter waveguide development, engineers considered using laser sources for systems of guided optical beams within pipes harboring lenses. But in 1966, Charles Kao had studied guided optical wave phenomena and predicted the emergence of low-loss optical fiber, a prediction that was fulfilled by a team of materials researchers at Corning in 1970. That development changed the world, and by the end of the century the vast majority of all long-distance transmission was over optical fibers at data rates approaching a terabit (a trillion bits per second) per fiber.

The Era of Information

The focus of research in communications prior to 1960 had been on the physical media for transmission. Little attention had been paid to the content of that transmission—the information being conveyed. But in the last half-century, three themes came together to focus attention on that informational content—the development of microelectronics, the shift to a digital representation of information, and the rise of an information economy.

The technological seeds of the Information Age were sown in a fertile period after the end of World War II. In 1947, the transistor was invented at Bell Labs by William Shockley, John Bardeen, and Walter Brattain. At the same time, John Mauchly and John Eckert were assembling 18,000 vacuum tubes into the ENIAC computer at the University of Pennsylvania. Meanwhile, Claude Shannon at Bell Labs was writing a landmark paper on a theory of information, and telecommunications engineers were just beginning to

ENIAC. Mammoth, clunky, and in constant need of human intervention, this famous vacuum-tube computer presaged a new era in both computation and communication. *University of Pennsylvania Archives.*

1981
IBM introduces the Personal Computer.

1983
Cellular phone network starts in U.S.

1989
Tim Berners-Lee and colleagues at the Swiss-based international elementary particle laboratory CERN create Hypertext Transfer Protocol (HTTP), a standardized communication mode for computer networks. The World Wide Web is launched.

1993
Mosaic, the first user-friendly graphical interface, is released; it greatly accelerates the proliferation of Web users.

1990s
The Internet rapidly becomes a socially transforming technology.

For a much more extensive timeline, see www.mediahistory.com/time/alltime.html.

consider the possible advantages of digital transmission using pulse code modulation, whose potential importance for encrypted transmission became apparent during the war.

At first the transistor was seen as a direct replacement for the vacuum tube, albeit smaller and requiring less power. But it became the first step on a pathway that constituted possibly the most significant technological development of the century. In 1958, Jack Kilby at Texas Instruments fabricated a chip with several transistors on a substrate, and the integrated circuit was born. Perhaps the transistors themselves were less important than the development of photolithographic fabrication, which enabled engineers to mass-produce on a microscale the wires and passive components that connected the active devices. Following Kilby's work was an inevitable evolution from calculator chips to microprocessors to personal computers. The availability, power, and digital form of this hardware enabled and shaped its applications in the subsequent Information Age.

In 1965 Gordon Moore, one of the founders of Intel and a pioneer of the Information Age, made an observation that has since become fundamental to business planning about technology evolution. This observation, called Moore's Law, states that there is a factor-of-two improvement in semiconductor technology every 18 months. This steady exponential improvement in the cost effectiveness of integrated circuits has maintained its pace almost exactly for the last three decades. Since 1965, this represents a gain of 2^{24}, or about eight orders of magnitude. Compare this with, for example, the difference between walking and flying in a jet plane, which is only two orders of magnitude. This unprecedented scaling of technology has made possible the computer and Internet technology on which science relies today.

It seems conceivable that technological progress has always been exponential, but that Moore's Law brought it to our attention because an exact measure of progress—the dimensions of circuit features—became possible with the advent of microelectronics. Progress in a number of related technical fields is also exponential, with various doubling periods. For example, optical capacity doubles every 12 months, Internet traffic doubles every 6 months, wireless capacity doubles every 9 months, and so forth.

The electronics revolution fostered the development of computers, the rise of computer networks, and the digitization of information and media. Together they created the present digital networked economy. It is hard to separate these themes and to say where one leaves off and the other begins; their evolution continues unabated. What's more, today's World Wide Web has been cited as a counterexample of the well-known thesis that all major technological developments require 25 years for widespread availability, as was the case with radio, television, and the telephone. The Web, by contrast, became overwhelmingly popular in only a few years. Of course, the Web needed the ubiquitous infrastructure of the Internet, which in turn required widespread availability of computers, which required the microprocessor, which required integrated circuits, and so forth. Certainly, it all goes back to the transistor, although it seems possible to make an argument that takes everything back to some development in antiquity. To paraphrase Newton, "We always are standing on the shoulders of giants."

In their influence on how science is transacted, the Internet and World Wide Web have had the greatest impact of any communications medium since possibly the printing press. The telegraph, telephone, and wireless were not different in concept from the postal system, except that the modern technologies were so much faster. The postal system, telephone, and telegraph are also one-to-one topologies, connecting a single user to another single, predesignated user. On the other hand, radio and television are one-to-many topologies for the broadcast of a small number of common channels to a great many receivers. The Internet and Web are something else entirely.

The beauty and power of these new media are that they allow the formation of spontaneous communities of unacquainted users. Their topology is neither one to one nor one to many, but rather many to many. They allow the sharing of information in textual, graphic, and multimedia formats across these communities, and they empower users within these communities to build their own applications. It is this empowerment of the periphery that has opened the floodgates of innovation to millions. In all previous communications technologies the ability to innovate and craft new systems and applications was confined to a small number of industrial engineers who tended the centralized intelligence and functionality.

The key idea of the Internet—a simple, common core protocol with intelligence at the periphery—was the critical ingredient of the culture from which the Internet arose. In the 1960s, computer communications centered upon the development of modems to enable shared access to expensive central computers over the existing telephone infrastructure, which was circuit-switched and designed entirely around voice transmission. Packet transmission and the open architecture that characterized the U.S. Defense Department's experimental network, ARPAnet, at its inception in 1969 had to come from outside the traditional industry. We are fortunate today that government and academia led this development. That's a big part of the reason why today's Internet is not fragmented into proprietary subnets but is instead open to all on an equal basis. It has been said that the greatest invention of the computer industry was not the PC, but rather the idea of an open platform that allows different innovators to mix and match their hardware and software. The Internet did the same thing for the telecommunications industry.

The protocol that defines the Internet, TCP/IP, was written by Robert Kahn and Vinton Cerf in 1973. Its genius, perhaps better understood in retrospect, was that it perfectly obeyed the maxim of being as simple as possible, but not more so. Consequently, the Internet is often viewed as an hourglass, with the multitude of physical media at the wide bottom and the plethora of applications at the wide top. The two are connected by the narrow neck of the Internet Protocol, IP, through which all communications must flow. It is a beautiful, flexible, and extensible architecture.

The most important application in the early days of the Internet turned out not to be access to time-shared computers but rather the simple e-mail that flowed between networked researchers. E-mail today remains a mainstay, but the World Wide Web became the sociotechnical invention that facilitated the kind of communications most important to science. It is perhaps not a coincidence that the Web came from a physics laboratory, CERN, where it was pioneered by Tim Berners-Lee in 1989. Subsequently, the first browser, Mosaic, was conceived at the National Center for Supercomputing Applications (NCSA) at the University of Illinois, Urbana-Champaign. This was followed by the commercial development of browsers and of search engines, which had originated at Carnegie-Mellon, Stanford, and Berkeley. All of the ingredients of an information environment to promote scientific collaboration fell into place.

Science and the Web Today

The infrastructure for research and collaboration in science today through the Internet and Web is rich with power and promise. Yet at the same time it is often filled with frustration. Who could have dreamed a decade ago that we would have instant access to a billion documents from the comfort and privacy of our office or laptop? What a pleasure it is to do library research today! However, we all frequently have the experience of starting a search for some topic, only to get sidetracked into interesting but irrelevant material, and never finding what we initially were seeking.

Imperfect as they are, it is a wonder that search engines exist at all. What a daunting job—crawling the Web daily to retrieve and inventory those billion pages! Using various relevance criteria, they reduce those pages by about two-thirds, but then they have to conduct user searches that are based on an average of only about two English words. It is not surprising that the links they return often overwhelm and undersatisfy the desires of the searcher. Some of the search improvements being pursued today include the use of information on context or popularity with other users, human categorization, and interactive search dialogs—improvements that are coming from the insight and innovation of scientists and engineers of many different disciplines.

The information access dream has often been stated as having the Library of Congress online. Unfortunately, the reality is very different. Web pages are full of junk and questionable material (as is the Library of Congress, for that matter) and often do not include the detailed material necessary for scientific research. Even though scientists are, as a group, entirely comfortable with computers and networks, they generally have been slow to provide online access to their journals.

One of the main obstacles has been the existing publication system, which depends on the sale of journals and magazines (particularly to libraries) to support its operation. Among Internet users a strong culture has evolved that believes that "information wants to be free." Like other businesses that rely on the sale and distribution of intellectual property—such as the publication, music, and movie industries—science has yet to evolve a satisfactory economic model that defines and protects the rights of intellectual property owners in the face of the perfect copying and widespread distribution enabled by digital networking technology.

Several other characteristics of the scientific establishment have hindered Web access to current research results. One is the need for peer review in order to establish some judgment on the material. In a world increasingly filled with questionable and irrelevant material, the guidance of peers regarding what is genuinely worth our time to read and examine has become more critical than ever. Even though Web publication can be nearly instantaneous, peer review still proceeds at a human pace. Another serious obstacle has been the tenure committees at universities and their reluctance to give full credit to online publication. In spite of these obstacles, there are a number of branches of science where fellow researchers exchange their latest research results—through listservs and other mechanisms—nearly in real time.

The Internet also is providing new mechanisms to enable scientists to collaborate at a distance. Programs that implement "whiteboards," where participants can sketch on a shared visual space, have been available for several years. Remote sharing and manipulation of instruments and lab facilities, such as telescopes and microscopes, is commonly done today. Videoconferencing over the Internet is relatively simple at low resolution, but higher resolutions and multiparty calls remain increasingly difficult with today's technology. Considerable work is being done to establish architectures and protocols for efficient video broadcast trees. Nonetheless, telepresence may always remain, in many respects, a poor substitute for traditional face-to-face interaction.

In the current sociology of the Net, the notion of portals is popular. These are single Web sites that serve as entrances to integrate material in a particular field. In science, a good example of taking this concept to the next level is the Biology Workbench at the NCSA (biology.ncsa.uiuc. edu). Web visitors to the Biology Workbench can access a number of worldwide biology databases and can choose among the collected application programs to perform operations on these databases. Users also can view their results in customized ways and are unaware of where the applications are actually running.

The Net is also increasingly important for continuing education, the lifeblood of the scientific profession. A number of courses are being put on the Web, and both universities and commercial enterprises are getting in the business of packaging educational material for distance learning. The advantages are ease of access and the ability to search and speed through material for the most efficient use of time. Typically, these Web-based courses offer a "talking head" (which enhances involvement, if nothing else), graphics for slides, and a moving transcript of the lecture.

It is worth reflecting upon the momentous improvements in the infrastructure for scientific work that have occurred in the last several decades because of information technology. The ability to simulate and graph results, to ask "what ifs" that might never be answerable with traditional lab-bench experimentation, to share results instantaneously, to search the world's information archives, and to be educated at one's own pace remotely are just a few of those improvements. What a pleasure it is to work in science today, and how far we have come from the days of libraries and slide rules!

Future Surprises

No futurist predicted the emergence of the World Wide Web. Yet in retrospect it seems to be an obvious idea. This epitomizes the way information technology evolves. Although the underlying technology trends—the exponential increases in processing power and bandwidth—are predictable, the applications are not. While the technologists speak of bandwidth and dream of video-intensive applications, society focuses on e-mail and the Web. The ways in which we use information processing and telecommunications appear to emerge from the chaos of the social milieu.

Technologically, we are headed to a time when bandwidth and processing will be unlimited and free. Latency will approach the speed-of-light delay, and even that will pose a problem for distributed computing applications. Service quality will approach the "five nines" (99.999%) availability of the traditional telephone network. Encryption will finally be freed of its political restraints, and security and privacy will be assured.

The way we tap into the network will also change. Much of the access to future online environments will be wireless and untethered, as will be our lifestyles. Except for voice conversations, which will be seen as just another application of the Net, most communication will be asynchronous, happening at the convenience of the users, who are increasingly turning into nomads who need to tap into media wherever they happen to be. But these technical problems are relatively simple compared with the sociotechnical engineering required to improve the three dimensions of communications—human to information, human to human, and human to computer.

In improving access to information, greater digitization of archived material and better search methodologies are necessary, and we can expect significant improvements in both areas. However, science has always been about the murky chain of converting data to information, to knowledge, and finally to wisdom. This process is likely to remain human-centered.

In the second dimension, human to human, the aim of communications has always been to replicate the experience of a face-to-face relationship at a distance. No communications technology has yet preserved the important nuances of a face-to-face interaction. There is, however, no reason to believe that perfect replication of such interactions is either achievable or desirable. Conceivably, by mediating the human interaction, communications at a distance could, in some ways, be better than face to face. A simple example would be simultaneous language translation, or the augmentation of the dialog with computer information retrieved in real time by agents that automatically roam the Web in search of relevant data. Nor are all the nuances of a face-to-face dialog helpful. For example, early experiments with videoconferencing showed that it was easier to deceive others using video than with audio alone—gestures and facial expressions could be used to draw attention away from lies or faulty logic.

Finally, communication between humans and computers can be improved by speech technology, natural language understanding, and machine implementation of commonsense reasoning. However, even though we have reached 2001, the year of HAL from Stanley Kubrick's famous movie, we are nowhere near to realizing HAL's capabilities for speech interaction. It will likely be some time before we start talking to computers like friends, and more time still before we think of computers as fellow scientists. Considering how HAL turned out in the movie, maybe this is just as well.

Further Reading

J. Abbate, *Inventing the Internet* (MIT Press, Cambridge, MA, 1999).

T. Berners-Lee, Mark Fischetti, Michael Dertouzos, *Weaving the Web: The Original Design and Ultimate Destiny of the World Wide Web by Its Inventor* (Harper San Francisco, 1999).

J. Brooks, *Telephone: The First Hundred Years* (Harper & Row, New York, NY, 1976).

R. W. Clark, *Einstein: The Life and Times* (Avon Books, New York, NY, reissue 1999).

G. B. Dyson, *Darwin Among the Machines: The Evolution of Global Intelligence* (Perseus Books, New York, NY, 1998).

8

One Hundred Years of Quantum Physics

DANIEL KLEPPNER AND ROMAN JACKIW

In the words of the authors of this essay, quantum theory is "the most precisely tested and most successful theory in the history of science." In just a few thousand words, they convey, in the most readable manner, a book's worth of information about how quantum theory came to be, how it changed the world, and how it might continue to evolve.

DANIEL KLEPPNER is Lester Wolfe Professor of Physics at MIT and Director of the Center for Ultracold Atoms. His research interests include atomic physics, quantum optics, ultraprecise spectroscopy, and Bose-Einstein condensation. ROMAN JACKIW is Jerrold Zacharias Professor of Physics at MIT. His research concerns application of quantum field theory to physical problems arising in particle, condensed matter, and gravitational physics.

Timeline of Quantum Physics

1897
Pieter Zeeman shows that light is radiated by the motion of charged particles in an atom, and Joseph John (J. J.) Thomson discovers the electron.

1900
Max Planck explains blackbody radiation in the context of quantized energy emission: Quantum theory is born.

1905
Albert Einstein proposes that light, which has wavelike properties, also consists of discrete, quantized bundles of energy, which are later called photons.

1911
Ernest Rutherford proposes the nuclear model of the atom.

1913
Niels Bohr proposes his planetary model of the atom, along with the concept of stationary energy states, and accounts for the spectrum of hydrogen.

An informed list of the most profound scientific developments of the twentieth century is likely to include general relativity, quantum mechanics, big bang cosmology, the unraveling of the genetic code, evolutionary biology, and perhaps a few other topics of the reader's choice. Among these, quantum mechanics is unique because of its profoundly radical quality. Quantum mechanics forced physicists to reshape their ideas of reality, to rethink the nature of things at the deepest level, and to revise their concepts of position and speed, as well as their notions of cause and effect.

Although quantum mechanics was created to describe an abstract atomic world far removed from daily experience, its impact on our daily lives could hardly be greater. The spectacular advances in chemistry, biology, and medicine—and in essentially every other science—could not have occurred without the tools that quantum mechanics made possible. Without quantum mechanics there would be no global economy, because the electronics revolution that brought us the computer age is a child of quantum mechanics. So is the photonics revolution that brought us the Information Age. The creation of quantum physics has transformed our world, bringing with it all the benefits—and the risks—of a scientific revolution.

Unlike general relativity, which grew out of a brilliant insight into the connection between gravity and geometry, or the deciphering of DNA, which unveiled a new world of biology, quantum mechanics did not spring from a single step. Rather, it was created in one of those rare concentrations of genius that occur from time to time in history. For 20 years after their introduction, quantum ideas were so confused that there was little basis for progress; then a small group of physicists created quantum mechanics in three tumultuous years. These scientists were troubled by what they were doing and in some cases distressed by what they had done.

The unique situation of this crucial yet elusive theory is perhaps best summarized by the following observation: Quantum theory is the most precisely tested and most successful theory in the history of science. Nevertheless, not only was quantum mechanics deeply disturbing to its founders, today—75 years after the theory was essentially cast in its current form—some of the luminaries of science remain dissatisfied with its foundations and its interpretation, even as they acknowledge its stunning power.

The year 2000 marked the 100th anniversary of Max Planck's creation of the quantum concept. In his seminal paper on thermal radiation, Planck hypothesized that the total energy of a vibrating system cannot be changed continuously. Instead, the energy must jump from one value to another in discrete steps, or quanta, of energy. The idea of energy quanta was so radical that

Papa Quanta. In 1900, Max Planck started the quantum-mechanical snowball. *Niels Bohr Library/American Institute of Physics.*

Planck let it lay fallow. Then Einstein, in his wonder year of 1905, recognized the implications of quantization for light. Even then the concept was so bizarre that there was little basis for progress. Twenty more years and a fresh generation of physicists were required to create modern quantum theory.

To understand the revolutionary impact of quantum physics one need only look at prequantum physics. From 1890 to 1900, physics journals were filled with papers on atomic spectra and essentially every other measurable property of matter, such as viscosity, elasticity, electrical and thermal conductivity, coefficients of expansion, indices of refraction, and thermoelastic coefficients. Spurred by the energy of the Victorian work ethic and the development of ever more ingenious experimental methods, knowledge accumulated at a prodigious rate.

What is most striking to the contemporary eye, however, is that the compendious descriptions of the properties of matter were essentially empirical. Thousands of pages of spectral data listed precise values for the wavelengths of the elements, but nobody knew why spectral lines occurred, much less what information they conveyed. Thermal and electrical conductivities were interpreted by suggestive models that fitted roughly half of the facts. There

1914
James Franck and Gustav Hertz confirm the existence of stationary states through an electron-scattering experiment.

1923
Arthur Compton observes that X rays behave like miniature billiard balls in their interactions with electrons, thereby providing further evidence for the particle nature of light.

1923
Louis de Broglie generalizes wave-particle duality by suggesting that particles of matter are also wavelike.

1924
Satyendra Nath Bose and Albert Einstein find a new way to count quantum particles, later called Bose-Einstein statistics, and they predict that extremely cold atoms should condense into a single quantum state, later known as a Bose-Einstein condensate.

1925
Wolfgang Pauli enunciates the exclusion principle.
Werner Heisenberg, Max Born, and Pascual Jordan develop matrix mechanics, the first version of quantum mechanics, and make an initial step toward quantum field theory.

1926
Erwin Schrödinger develops a second description of quantum physics, called wave mechanics. It includes what becomes one of the most famous formulae of science, which is later known as the Schrödinger equation.
Enrico Fermi and Paul A. M. Dirac find that quantum mechanics requires a second way to count particles, Fermi-Dirac statistics, opening the way to solid-state physics.
Dirac publishes a seminal paper on the quantum theory of light.

1927
Heisenberg states his Uncertainty Principle, that it is impossible to exactly measure both the position and momentum of a particle at the same time.

were numerous empirical laws, but they were not satisfying. For instance, the Dulong-Petit law established a simple relation between specific heat and the atomic weight of a material. Much of the time it worked; sometimes it did not. The masses of equal volumes of gas were in the ratios of integers—mostly. The Periodic Table, which provided a key organizing principle for the flourishing science of chemistry, had absolutely no theoretical basis.

Among the greatest achievements of the revolution is this: Quantum mechanics has provided a quantitative theory of matter. We now understand essentially every detail of atomic structure; the Periodic Table has a simple and natural explanation; and the vast arrays of spectral data fit into an elegant theoretical framework. Quantum theory permits the quantitative understanding of molecules, of solids and liquids, and of conductors and semiconductors. It explains bizarre phenomena such as superconductivity and superfluidity, and exotic forms of matter such as the stuff of neutron stars and Bose-Einstein condensates, in which all the atoms in a gas behave like a single superatom. Quantum mechanics provides essential tools for all of the sciences and for every advanced technology.

Quantum physics actually encompasses two entities. The first is the theory of matter at the atomic level: quantum mechanics. It is quantum mechanics that allows us to understand and manipulate the material world. The second is the quantum theory of fields. Quantum field theory plays a totally different role in science, to which we shall return later.

Quantum Mechanics

The clue that triggered the quantum revolution came not from studies of matter but from a problem in radiation. The specific challenge was to understand the spectrum of light emitted by hot bodies: blackbody radiation. The phenomenon is familiar to anyone who has stared at a fire. Hot matter glows, and the hotter it becomes the brighter it glows. The spectrum of the light is broad, with a peak that shifts from red to yellow and finally to blue (although we cannot see that) as the temperature is raised.

It should have been possible to understand the shape of the spectrum by combining concepts from thermodynamics and electromagnetic theory, but all attempts failed. However, by assuming that the energies of the vibrating electrons that radiate the light are quantized, Planck obtained an expression that agreed beautifully with experiment. But as he recognized all too well, the theory was physically absurd, "an act of desperation," as he later described it.

Planck applied his quantum hypothesis to the energy of the vibrators in the walls of a radiating body. Quantum physics might have ended there if in 1905 a novice—Albert Einstein—had not reluctantly concluded that if a vibrator's energy is quantized, then the energy of the electromagnetic field that it radiates—light—must also be quantized. Einstein thus imbued light with particlelike behavior, notwithstanding that James Clerk Maxwell's theory, and over a century of definitive experiments, testified to light's wave nature. Experiments on the photoelectric effect in the following decade revealed that when light is absorbed its energy actually arrives in discrete bundles, as if carried by a particle. The dual nature of light—particlelike or wavelike depending on what one looks for—was the first example of a vexing theme that would recur throughout quantum physics. The duality constituted a theoretical conundrum for the next 20 years.

The first step toward quantum theory had been precipitated by a dilemma about radiation. The second step was precipitated by a dilemma about matter. It was known that atoms contain positively and negatively charged particles. But oppositely charged particles attract. According to electromagnetic theory, therefore, they should spiral into each other, radiating light in a broad spectrum until they collapse.

Once again, the door to progress was opened by a novice: Niels Bohr. In 1913, Bohr proposed a radical hypothesis: Electrons in an atom exist only in certain stationary states, including a ground state. Electrons change their energy by "jumping" between the stationary states, emitting light whose wavelength depends on the energy difference. By combining known laws with bizarre assumptions about quantum behavior, Bohr swept away the problem of atomic stability. Bohr's theory was full of contradictions, but it provided a quantitative description of

Atoms go quantum. In 1913, Niels Bohr ushered quantum physics into world of atoms. *Niels Bohr Library/American Institute of Physics.*

1928
Dirac presents a relativistic theory of the electron that includes the prediction of antimatter.

1932
Carl David Anderson discovers antimatter, an anti-electron called the positron.

1934
Hideki Yukawa proposes that nuclear forces are mediated by massive particles called mesons, which are analogous to the photon in mediating electromagnetic forces.

1946–48
Experiments by Isidor I. Rabi, Willis Lamb, and Polykarp Kusch reveal discrepancies in the Dirac theory.

1948
Richard Feynman, Julian Schwinger, and Sin-Itiro Tomonaga develop the first complete theory of the interaction of photons and electrons, quantum electrodynamics, which accounts for the discrepancies in the Dirac theory.

the spectrum of the hydrogen atom. He recognized both the success and the shortcomings of his model. With uncanny foresight, he rallied physicists to create a new physics. His vision was eventually fulfilled, although it took 12 years and a new generation of young physicists.

At first, attempts to advance Bohr's quantum ideas—the so-called old quantum theory—suffered one defeat after another. Then a series of developments totally changed the course of thinking.

In 1923 Louis de Broglie, in his Ph.D. thesis, proposed that the particle behavior of light should have its counterpart in the wave behavior of particles. He associated a wavelength with the momentum of a particle: The higher the momentum the shorter the wavelength. The idea was intriguing, but no one knew what a particle's wave nature might signify or how it related to atomic structure. Nevertheless, de Broglie's hypothesis was an important precursor for events soon to take place.

Getting weirder. Louis de Broglie said that if wave-like light can behave like particles, then particles can behave like waves. *Niels Bohr Library/American Institute of Physics.*

In the summer of 1924, there was yet another precursor. Satyendra N. Bose proposed a totally new way to explain the Planck radiation law. He treated light as if it were a gas of massless particles (now called photons) that do not obey the classical laws of Boltzmann statistics but behave according to a new type of statistics based on particles' indistinguishable nature. Einstein immediately applied Bose's reasoning to a real gas of massive particles and obtained a new law—to become known as the Bose-Einstein distribution—for how energy is shared by the particles in a gas. Under normal circumstances, however, the new and old theories predicted the same behavior for atoms in a gas. Einstein took no further interest, and the result lay undeveloped for more than a decade. Still, its key idea, the indistinguishability of particles, was about to become critically important.

Suddenly, a tumultuous series of events occurred, culminating in a scientific revolution. In the three-year period from January 1925 to January 1928:

- Wolfgang Pauli proposed the exclusion principle, providing a theoretical basis for the Periodic Table.
- Werner Heisenberg, with Max Born and Pascual Jordan, discovered matrix mechanics, the first version of quantum mechanics. The historical

1957
John Bardeen, Leon Cooper, and J. Robert Schrieffer show that electrons can form pairs whose quantum properties allow them to travel without resistance, providing an explanation for the zero electrical resistance of superconductors.

1959
Yakir Aharonov and David Bohm predict that a magnetic field affects the quantum properties of an electron in a way that is forbidden by classical physics. The

goal of understanding electron motion within atoms was abandoned in favor of a systematic method for organizing observable spectral lines.

- Erwin Schrödinger invented wave mechanics, a second form of quantum mechanics in which the state of a system is described by a wave function, the solution to Schrödinger's equation. Matrix mechanics and wave mechanics, apparently incompatible, were shown to be equivalent.
- Electrons were shown to obey a new type of statistical law, Fermi-Dirac statistics. It was recognized that all particles obey either Fermi-Dirac statistics or Bose-Einstein statistics, and that the two classes have fundamentally different properties.
- Heisenberg enunciated the Uncertainty Principle.
- Paul A. M. Dirac developed a relativistic wave equation for the electron that explained electron spin and predicted antimatter.
- Dirac laid the foundations of quantum field theory by providing a quantum description of the electromagnetic field.
- Bohr announced the complementarity principle, a philosophical principle that helped to resolve apparent paradoxes of quantum theory, particularly wave-particle duality.

The principal players in the creation of quantum theory were young. In 1925, Pauli was 25 years old, Heisenberg and Enrico Fermi were 24, and Dirac and Jordan were 23. Schrödinger, at age 36, was a late bloomer. Born and Bohr were older still, and it is significant that their contributions were largely interpretative. The profoundly radical nature of the intellectual achievement is revealed by Einstein's reaction. Having invented some

Unknowable reality. Werner Heisenberg articulated one of the most societally absorbed ideas of quantum physics: the Uncertainty Principle. *Niels Bohr Library/American Institute of Physics.*

Omniscient math. It's tough to solve, but Erwin Scrödinger's famous equation describes every observable state of a physical system. *Niels Bohr Library/American Institute of Physics.*

Aharonov-Bohm effect is observed in 1960 and hints at a wealth of unexpected macroscopic effects.

1960
Building on work by Charles Townes, Arthur Schawlow, and others, Theodore Maiman builds the first practical laser.

1964
John S. Bell proposes an experimental test, "Bell's inequalities," of whether quantum mechanics provides the most complete possible description of a system.

1970s
Foundations are laid for the Standard Model of Particle Physics, in which matter is said to be built of quarks and leptons that interact via the four physical forces.

1982
Alain Aspect carries out an experimental test of Bell's inequalities and confirms the completeness of quantum mechanics.

1995

Eric Cornell, Carl Wieman, and Wolfgang Ketterle trap clouds of metallic atoms cooled to less than a millionth of a degree above absolute zero, producing Bose–Einstein condensates, which were first predicted 70 years earlier. This accomplishment leads to the creation of the atom laser and superfluid gases.

For more extensive timelines of quantum physics, see two of Abraham Pais's books: *Inward Bound: Of Matter and Forces in the Physical World* and *Niels Bohr's Times: In Physics, Philosophy, and Polity*. See also timeline.aps.org/APS/index.html.

of the key concepts that led to quantum theory, Einstein rejected it. His paper on Bose-Einstein statistics was his last contribution to quantum physics and his last significant contribution to physics.

That a new generation of physicists was needed to create quantum mechanics is hardly surprising. Lord Rayleigh described why in a letter to Bohr congratulating him on his 1913 paper on hydrogen. He said that there was much truth in Bohr's paper, but he would never understand it himself. Rayleigh recognized that radically new physics would need to come from unfettered minds.

In 1928, the revolution was finished and the foundations of quantum mechanics were essentially complete. The frenetic pace with which it occurred is revealed by an anecdote recounted by the late Abraham Pais in *Inward Bound*. In 1925, the concept of electron spin had been proposed by Samuel Goudsmit and George Uhlenbeck. Bohr was deeply skeptical. In December, he traveled to Leiden, the Netherlands, to attend the jubilee of Hendrik A. Lorentz's doctorate. Pauli met the train at Hamburg, Germany, to find out Bohr's opinion about the possibility of electron spin. Bohr said the proposal was "very, very interesting," his well-known put-down phrase. Later at Leiden, Einstein and Paul Ehrenfest met Bohr's train, also to discuss spin. There, Bohr explained his objection, but Einstein showed a way around it and converted Bohr into a supporter. On his return journey, Bohr met with yet more discussants. When the train passed through Göttingen, Germany, Heisenberg and Jordan were waiting at the station to ask his opinion. And at the Berlin station, Pauli was waiting, having traveled especially from Hamburg. Bohr told them all that the discovery of electron spin was a great advance.

The creation of quantum mechanics triggered a scientific gold rush. Among the early achievements were these: Heisenberg laid the foundations for atomic structure theory by obtaining an approximate solution to Schrödinger's equation for the helium atom in 1927, and general techniques for calculating the structures of atoms were created soon after by John Slater, Douglas Rayner Hartree, and Vladimir Fock. The structure of the hydrogen molecule was solved by Fritz London and Walter Heitler; Linus Pauling built on their results to found theoretical chemistry. Arnold Sommerfeld and Pauli laid the foundations of the theory of electrons in metals, and Felix Bloch created band structure theory. Heisenberg explained the origin of ferromagnetism. The enigma of the random nature of radioactive decay by alpha particle emission was explained in 1928 by George Gamow, who showed that it occurs by quantum-mechanical tunneling. In the following years Hans Bethe laid the foundations for nuclear physics and explained the energy source of stars. With these developments atomic, molecular, solid state, and nuclear physics entered the modern age.

Controversy and Confusion

Alongside these advances, however, fierce debates were taking place on the interpretation and validity of quantum mechanics. Foremost among the protagonists were Bohr and Heisenberg, who embraced the new theory, and Einstein and Schrödinger, who were dissatisfied. To appreciate the reasons for such turmoil, one needs to understand some of the key features of quantum theory, which we summarize here. (For simplicity, we describe the Schrödinger version of quantum mechanics, sometimes called wave mechanics.)

Fundamental description: the wave function.

The behavior of a system is described by Schrödinger's equation. The solutions to Schrödinger's equation are known as wave functions. The complete knowledge of a system is described by its wave function, and from the wave function one can calculate the possible values of every observable quantity. The probability of finding an electron in a given volume of space is proportional to the square of the magnitude of the wave function. Consequently, the location of the particle is "spread out" over the volume of the wave function. The momentum of a particle depends on the slope of the wave function: The greater the slope, the higher the momentum. Because the slope varies from place to place, momentum is also "spread out." The need to abandon a classical picture in which position and velocity can be determined with arbitrary accuracy in favor of a blurred picture of probabilities is at the heart of quantum mechanics.

Measurements made on identical systems that are identically prepared will not yield identical results. Rather, the results will be scattered over a range described by the wave function. Consequently, the concept of an electron having a particular location and a particular momentum loses its foundation. The Uncertainty Principle quantifies this: To locate a particle precisely, the wave function must be sharply peaked (that is, not spread out). However, a sharp peak requires a steep slope, and so the spread in momentum will be great. Conversely, if the momentum has a small spread, the slope of the wave function must be small, which means that it must spread out over a large volume, thereby portraying the particle's location less exactly.

Waves can interfere.

Their heights add or subtract depending on their relative phase. Where the amplitudes are in phase, they add; where they are out of phase, they subtract. If a wave can follow several paths from source to receiver, as a light wave undergoing two-slit interference, then the illumination will generally display interference fringes. Particles obeying a wave equation will do likewise, as in electron diffraction. The analogy seems reasonable until one inquires about the nature of the wave. A wave is generally thought of as a disturbance in a medium. In quantum mechanics there is no medium, and in a sense there is no wave, as the wave function is fundamentally a statement of our knowledge of a system.

Symmetry and identity.

A helium atom consists of a nucleus surrounded by two electrons. The wave function of helium describes the position of each electron. However, there is no way of distinguishing which electron is which. Consequently, if the electrons are switched the system must look the same, which is to say the probability of finding the electrons in given positions is unchanged. Because the probability depends on the square of the magnitude of the wave function, the wave function for the system with the interchanged particles must be related to the original wave function in one of two ways: Either it is identical to the original wave function, or its sign is simply reversed, i.e., it is multiplied by a factor of –1. Which one is it?

One of the astonishing discoveries in quantum mechanics is that for electrons the wave function always changes sign. The consequences are dramatic, for if two electrons are in the same quantum state, then the wave function has to be its negative opposite. Consequently, the wave function must vanish. Thus, the probability of finding two electrons in the same state is zero. This is the Pauli exclusion principle. All particles with half-integer spin, including electrons, behave this way and are called fermions. For particles with integer spin, including photons, the wave function does not change sign. Such particles are called bosons. Electrons in an atom arrange themselves in shells because they are fermions, but light from a laser emerges in a single superintense beam—essentially a single quantum state—because light is composed of bosons. Recently, atoms in a gas have been cooled to the quantum regime where they form a Bose-Einstein condensate, in which the system can emit a superintense matter beam—forming an atom laser.

These ideas apply only to identical particles, because if different particles are interchanged the wave function will certainly be different. Consequently, particles behave like fermions or like bosons only if they are *totally* identical. The absolute identity of like particles is among the most mysterious aspects of quantum mechanics. Among the achievements of quantum field theory is that it can explain this mystery.

What does it mean?

Questions such as what a wave function "really is" and what is meant by "making a measurement" were intensely debated in the early years. By 1930, however, a more or less standard interpretation of quantum mechanics had been developed by Bohr and his colleagues, the so-called Copenhagen interpretation. The key elements are the probabilistic description of matter and events, and reconciliation of the wavelike and particlelike natures of things through Bohr's principle of complementarity. Einstein never accepted quantum theory. He and Bohr debated its principles until Einstein's death in 1955.

A central issue in the debates on quantum mechanics was whether the wave function contains all possible information about a system or if there might be underlying factors—hidden variables—that determine the outcome of a particular measurement. In the mid-1960s John S. Bell showed that if hidden variables existed, experimentally observed probabilities would have to fall below certain limits, dubbed "Bell's inequalities." Experiments were carried out by a number of groups, which found that the inequalities were violated. Their collective data came down decisively against the possibility of hidden variables. For most scientists, this resolved any doubt about the validity of quantum mechanics.

Nevertheless, the nature of quantum theory continues to attract attention because of the fascination with what is sometimes described as "quantum weirdness." The weird properties of quantum systems arise from what is known as entanglement. Briefly, a quantum system, such as an atom, can exist in any one of a number of stationary states but also in a superposition—or sum—of such states. If one measures some property such as the energy of an atom in a superposition state, in general the result is sometimes one value, sometimes another. So far, nothing is weird.

It is also possible, however, to construct a two-atom system in an entangled state in which the properties of both atoms are shared with each other. If the atoms are separated, information about one is shared, or entangled, in the state of the other. The behavior is unexplainable except in the language of quantum mechanics. The effects are so surprising that they are the focus of study by a small but active theoretical and experimental community. The issues are not limited to questions of principle, as entanglement can be useful. Entangled states have already been employed in quantum communication systems, and entanglement underlies all proposals for quantum computation.

The Second Revolution

During the frenetic years in the mid-1920s when quantum mechanics was being invented, another revolution was under way. The foundations were being laid for the second branch of quantum physics—quantum field theory. Unlike quantum mechanics, which was created in a short flurry of activity and emerged essentially complete, quantum field theory has a tortuous history that continues today. In spite of the difficulties, the predictions of quantum field theory are the most precise in all of physics, and quantum field theory constitutes a paradigm for some of the most crucial areas of theoretical inquiry.

The problem that motivated quantum field theory was the question of how an atom radiates light as its electrons "jump" from excited states to the ground state. Einstein proposed such a process, called spontaneous emission, in 1916, but he had no way to calculate its rate. Solving the problem required developing a fully relativistic quantum theory of electromagnetic fields, a quantum theory of light. Quantum mechanics is the theory of matter. Quantum field theory, as its name suggests, is the theory of fields, not only electromagnetic fields but other fields that were subsequently discovered.

In 1925 Born, Heisenberg, and Jordan published some initial ideas for a theory of light, but the seminal steps were taken by Dirac—a young and essentially unknown physicist working in isolation—who presented his field theory in 1926. The theory was full of pitfalls: formidable calculational complexity, predictions of infinite quantities, and apparent violations of the correspondence principle.

In the late 1940s a new approach to the quantum theory of fields, QED (for quantum electrodynamics), was developed by

Fields go quantum. Paul Dirac spearheaded work leading to quantum field theory as well as discoveries such as antimatter. *Niels Bohr Library/American Institute of Physics,*

Richard Feynman, Julian Schwinger, and Sin-Itiro Tomonaga. They sidestepped the infinities by a procedure, called renormalization, which essentially subtracts infinite quantities so as to leave finite results. Because there is no exact solution to the complicated equations of the theory, an approximate answer is presented as a series of terms that become more and more difficult to calculate. Although the terms become successively smaller, at some point they should start to grow, indicating the breakdown of the approximation. In spite of these perils, QED ranks among the most brilliant successes in the history of physics. Its prediction of the interaction strength between an electron and a magnetic field has been experimentally confirmed to a precision of two parts in 1,000,000,000,000.

Notwithstanding its fantastic successes, QED harbors enigmas. The view of empty space—the vacuum—that the theory provides initially seems preposterous. It turns out that empty space is not really empty. Rather, it is filled with small, fluctuating electromagnetic fields. These vacuum fluctuations are essential for explaining spontaneous emission. Furthermore, they produce small but measurable shifts in the energies of atoms and certain properties of particles such as the electron. Strange as they seem, these effects have been confirmed by some of the most precise experiments ever carried out.

At the low energies of the world around us, quantum mechanics is fantastically accurate. But at high energies where relativistic effects come into play, a more general approach is needed. Quantum field theory was invented to reconcile quantum mechanics with special relativity.

The towering role that quantum field theory plays in physics arises from the answers it provides to some of the most profound questions about the nature of matter. Quantum field theory explains why there are two fundamental classes of particles—fermions and bosons—and how their properties are related to their intrinsic spin. It describes how particles—not only photons, but electrons and positrons (antielectrons)—are created and annihilated. It explains the mysterious nature of identity in quantum mechanics—how identical particles are absolutely identical because they are created by the same underlying field. QED describes not only the electron but the class of particles called leptons that includes the muon, the tau meson, and their antiparticles.

Because QED is a theory for leptons, however, it cannot describe more complex particles called hadrons. These include protons, neutrons, and a wealth of mesons. For hadrons, a new theory had to be invented, a generalization of QED called quantum chromodynamics, or QCD. Analogies abound between QED and QCD. Electrons are the constituents of atoms; quarks are the constituents of hadrons. In QED the force between charged particles is mediated by the photon; in QCD the force between quarks is mediated by the gluon. In spite of the parallels, there is a crucial difference between QED and QCD. Unlike leptons and photons, quarks and gluons are forever confined within the hadron. They cannot be liberated and studied in isolation.

QED and QCD are the cornerstones for a grand synthesis known as the Standard Model. The Standard Model has successfully accounted for every particle experiment carried out to date. However, for many physicists the Standard Model is inadequate, because data on the masses, charges, and other properties of the fundamental particles need to be found from experiments. An ideal theory would predict all of these.

Today, the quest to understand the ultimate nature of matter is the focus of an intense scientific study that is reminiscent of the frenzied and miraculous days in which quantum mechanics was created, and whose outcome may be even more far-reaching. The effort is inextricably bound to the quest for a quantum description of gravity. The procedure for quantizing the electromagnetic field that worked so brilliantly in QED has failed to work for gravity, in spite of a half-century of effort. The problem is critical, for if general relativity and quantum mechanics are both correct, then they must ultimately provide a consistent description for the same events. There is no contradiction in the normal world around us, because gravity is so fantastically weak compared to the electrical forces in atoms that quantum effects of gravity are negligible and a classical description works beautifully. But for a system such as a black hole where gravity is incredibly strong, we have no reliable way to predict quantum behavior.

One century ago our understanding of the physical world was empirical. Quantum physics gave us a theory of matter and fields, and that knowledge transformed our world. Looking to the next century, quantum mechanics will continue to provide fundamental concepts and essential tools for all of the sciences. We can make such a prediction confidently because for the world around us quantum physics provides an exact and complete theory. However, physics today has this in common with physics in 1900: It remains ultimately empirical—we cannot fully predict the properties of the elementary constituents of matter, we must measure them.

Perhaps string theory—a generalization of quantum field theory that eliminates all infinities by replacing pointlike objects such as the electron with extended objects—or some theory only now being conceived will solve the riddle. Whatever the outcome, the dream of ultimate understanding will continue to be a driving force for new knowledge, as it has been since the dawn of science. One century from now, the consequences of pursuing that dream will belie our imagination.

Further Reading

B. Bederson, Ed., More *Things in Heaven and Earth: A Celebration of Physics at the Millennium* (Springer-Verlag, New York, 1999).

J. S. Bell, *Speakable and Unspeakable in Quantum Mechanics: Collected Papers on Quantum Mechanics* (reprint edition) (Cambridge University Press, Cambridge, 1989).

L. M. Brown, A. Pais, B. Pippard, Eds., *Twentieth Century Physics* (Institute of Physics, Philadelphia, 1995).

D. Cassidy, Uncertainty: *The Life and Science of Werner Heisenberg* (W. H. Freeman, New York, 1993).

A. Einstein, *Born-Einstein Letters,* trans. Irene Born (Macmillan, London, 1971).

H. Kragh, *Dirac: A Scientific Biography* (Cambridge University Press, Cambridge, 1990).

W. Moore, *Schrödinger: Life and Thought* (Cambridge University Press, Cambridge, 1989).

A. Pais, *Inward Bound: Of Matter and Forces in the Physical World* (Oxford University Press, Oxford, 1986).

A. Pais, *Niels Bohr's Times: In Physics, Philosophy, and Polity* (Oxford University Press, Oxford, 1991).

9

The Incredible Life and Times of Biological Cells

PAUL NURSE

In an essay on the science of cells that bridges Anne McLaren's essay on cloning and Eric Lander and Robert Weinberg's essay on genomics, Paul Nurse recapitulates the ontogeny of one of the most important theories in the history of biology, the cell theory. Nurse then lays out the wondrous molecular complexities and processes that he and others have discovered while studying the lives of cells. As the vast diversity, interconnectedness, and elegance of the life-sustaining molecular choreography inside cells becomes more visible, an adjective like *miraculous* ceases to fully describe what we've learned.

PAUL NURSE is director general of the Imperial Cancer Research Fund (ICRF) in London, where he previously had served as director of research. He also is head of ICRF's Cell Cycle Laboratory, a research group that studies the genes that prompt cells to divide. He is well known for his contributions to the discovery of the molecular mechanisms that orchestrate cell divisions in most living organisms. In 2001, Nurse was named a co-recipient of the Nobel Prize in Physiology or Medicine for his work on the cell cycle.

Timeline of Cell Biology

1609
Hans Lippershey and Zacharias Janssen independently invent the compound microscope, which later enables scientists to visualize cells.

1665
Robert Hooke publishes *Micrographia,* in which he describes cells for the first time based on his observations of cork sections.

1675
Marcello Malpighi publishes *Anatome Plantarum,* the first important work on plant anatomy based on microscopic observation.

1676
Using a simple microscope and pond scum, Anton van Leeuwenhoek observes tiny organisms, or what he calls animalcules.

One of the most important ideas in the history of biology is the cell theory, which proposes that all forms of life are composed of cells. Cells are the simplest units to exhibit the functions characteristic of life, and the field of cell biology has helped reveal how the immense variety of life forms are organized and operate. In this chapter, I trace the pathways that led to the ascent of the cell to its central role in biological understanding. Along the way, I describe discoveries that identified the cell as the fundamental structural and functional unit of life. I focus particularly on the mechanisms and controls of cell reproduction that ultimately allow growth, development, and evolution to occur. Finally, I will speculate about how future discoveries should provide further understanding of how cells function.

Finding Life's Pixel [1–4]

As is often the case in science, technology begets discovery. The invention of the microscope led to the discovery of the cell. Improvements in lenses in early seventeenth-century Holland resulted in the construction of simple microscopes that were used to investigate insects and other small organisms. Using a compound microscope, Robert Hooke extended this work to examine sections of cork. He published drawings of what he saw in his 1665 book *Micrographia,* in which he clearly illustrated the walled cavities of what he termed "cells," after the Latin *cella*, for small room or cubicle.

Within a few years, plant cells had been comprehensively characterized in two beautifully illustrated studies, one by Nehemiah Grew and one by Marcello Malpighi. Their work led to the view that plant tissues are mostly com-

Hooke looks. Seeing cells with his compound microscope, Robert Hooke was able to discern the cellular structure of cork. *Dover Publications.*

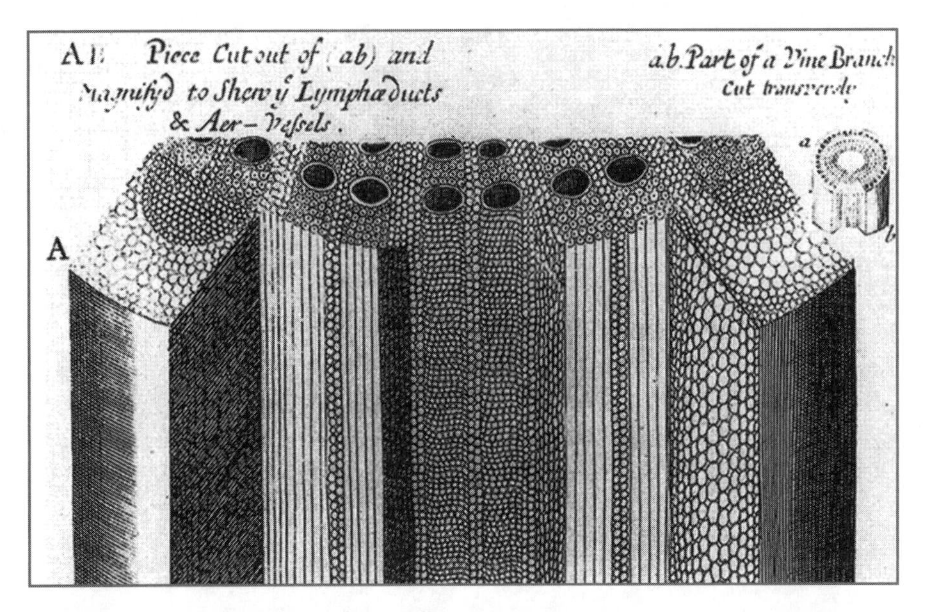

Cells galore. Drawings, like this one of a vine stem section by Nehemiah Grew in 1682, bolstered the idea that cells make up biological tissue. *Harris, The Birth of the Cell.*

posed of aggregates of cells. A little later in the century, Malpighi, Anton van Leeuwenhoek, and Jan Swammerdam became the first scientists to recognize cells in animals. These microscopists described corpuscles in blood, but no one at this time proposed that solid animal tissues also were made up of cells. The oversight is understandable. Animal tissues were more difficult to preserve than plant tissues, and they presented a more fibrous appearance compared to the well-defined cellular geometry of plant tissues. Leeuwenhoek also discovered single-celled organisms, which he called animalcules, growing in extracts of plants. Improvements in microscopical observation led to better descriptions of cells, and in 1766 Abraham Trembley reported observing animalcules (the protozoan synhedra) reproducing; it probably was the first report of binary fission of a cell.

The gradual acceptance of the atomic nature of matter also helped in the development of the cell theory. The idea that all matter might be composed of indivisible subunits, or atoms, arose in the fifth century B.C. in Greece, but it took two millennia for atomism to become a serious scientific topic. By the eighteenth century, it had become natural for biologists to think about fundamental subunits of living matter when interpreting the microscopic structures of plants and animals. A pivotal speculation was made by Lorenz Oken

1682

Nehemiah Grew publishes *The Anatomy of Plants,* in which he shows that plant tissues are composed mostly of small chambers or cells.

Late seventeenth century

Leeuwenhoek, Malpighi, and Jan Swammerdam describe corpuscles in blood.

1766

In a letter, Abraham Trembley writes what is probably the first description of binary fission of a cell.

1805

Lorenz Oken argues that plants and animals are assemblages of "infusoria"— microbes such as protozoa that scientists had been observing in extracts, or infusions, of plant and animal tissues.

1812

Jöns Berzelius coins the term "catalysis" to describe reactions in which certain substances participate yet are not consumed.

1828
Friedrich Wöhler synthesizes urea in test tubes, thereby disproving the notion that a vital principle is required for the creation of the molecules of living things.

1830s
Jan Purkinje and Gabriel Valentin note that animal tissues, like those of plants, are composed of cells.

1831
Robert Brown identifies the small body found within plant cells as the "nucleus."

1835
Charles Cagniard-Latour shows that fermentation reactions are associated with living organisms.

1838–39
Theodor Schwann and Matthias Schleiden consolidate many observations to develop what becomes known as the cell theory of biology.

1856
Louis Pasteur reports that fermentation is caused by microorganisms.

Going cellular. In 1858, Rudolf Virchow published *Cellularpathologie,* in which he states *"Omnis cellula e cellula"*; that is, all cells come from cells. *National Library of Medicine.*

in 1805. He argued that multicellular plants and animals are assemblages of the tiny living "infusoria," such as protozoa, that grow in animal and plant extracts. In succeeding years, microscopists generalized that idea by noting parallelisms between infusoria, on the one hand, and animal and plant cells, on the other.

These studies culminated in the cell theory. It was popularized by the botanist Matthias Schleiden and the zoologist Theodor Schwann, who in 1839 stated, "We have seen that all organisms are composed of essentially like parts, namely of cells." In his 1858 book *Cellularpathologie,* Rudolf Virchow stated, "Every animal appears as a sum of vital units, each of which bears in itself the complete characteristics of life." This discovery—that cells form the fundamental structural and functional units of all living organisms—was a landmark in the history of biology.

Cells Do It All: Heredity, Development, Disease, and Death [1,4]

Although Schleiden and Schwann correctly articulated the cell theory, they were confused about the formation of cells, thinking that they arose by processes akin to precipitation or crystallization. Others, particularly Barthelemy Dumortier working with plant cells and Robert Remak with animal cells, recognized that cells arose from preexisting cells by a process of binary fission. This view was well championed by Virchow, who popularized the phrase *"Omnis cellula e cellula,"* that is, all cells come from cells.

Research into cell reproduction began revealing the cell's wondrous complexity and machinelike features. At the heart of cellular reproduction is a system of heredity, which works by a molecular mechanism that ultimately explained Mendel's laws of inheritance. In the 1860s, Gregor Mendel had concluded that attributes of plants, such as seed shape and stem height, were determined by pairs of characters, one derived from the male plant and the other from the female. He thought these characters remained distinct in the hybrid plant and were passed on as discrete entities in further crosses. These conclusions became more scientifically acceptable only at the turn of the twentieth century, when researchers found that the behavior of Mendel's characters mirrored the behavior of chromosomes at cell division.

Most cells contain a single nucleus that reproduces during mitosis and cell division. Elongated chromosomal threads, which Walther Flemming and Eduard Strasburger described in the 1880s, were observed to split lengthways before shortening and thickening as mitosis proceeds. The longitudinal halves then separate into the two daughter nuclei. About the same time as these observations were made, Edouard van Beneden showed that the chromosomes in a fertilized nematode egg are derived in equal numbers from the egg and the sperm. That led August Weisman to propose that the heredity system is based on the chromosomes.*

Even as scientists were developing their first ideas about the mechanisms behind cell reproduction, they were discerning connections between such reproduction and the growth and development of organisms. Population growth of single-celled microbial organisms entails cell division, and by the late nineteenth century researchers understood that the growth of multicellular organisms is also based on cell division. In the 1860s, Rudolf Kölliker observed that the cleavage of early embryos reflects a series of cell divisions yielding cells that subsequently differentiate into specialized tissues. And by the 1880s, scientists had concluded that all multicellular organisms, regardless of their complexity, emerge from a single cell.

*The mapping of genes to chromosomes, as well as the discoveries that genes are made of DNA and encode the proteins that determine the characteristics of cells and organisms, will not be considered further here, as they are central themes of a previous chapter, "Genomics: Journey to the Center of Biology."

1858
Rudolf Virchow publishes *Cellularpathologie,* in which he states *"Omnis cellula e cellula"*; that is, all cells come from cells.

1859
Charles Darwin publishes his watershed book on evolution by natural selection, known usually as *The Origin of Species.*

Early 1860s
Gregor Mendel discovers the laws of heredity.

1861
Rudolf Kölliker and others begin interpreting embryology in terms of cell theory.

1882
Walther Flemming and Eduard Strasburger describe elongated chromosomal threads forming from the nucleus during the onset of mitosis.

1892
August Weismann proposes that chromosomes are at the heart of heredity.

1897

Eduard and Hans Buchner show that a yeast extract containing a catalytic substance, zymase, promotes fermentation reactions. This evolves into a cornerstone of biochemistry, namely, that protein catalysts, or enzymes, promote life's chemistry.

1912

Jacques Loeb publishes his essay collection, *The Mechanistic Conception of Life,* which champions the view of cells as chemical machines.

1925

Edmund B. Wilson writes that "The key to every biological problem must finally be sought in the cell."

1945

The first electron micrograph of an intact cell is published in *The Journal of Experimental Medicine* by Keith Porter, Albert Claude, and Ernest Fullam.

1950s

Researchers discover S-phase (during which DNA synthesis occurs), which allows the cell cycle to be divided into four phases known as G_1, S, G_2, and M.

That cell reproduction is central to growth and development makes immediate sense. That the death of cells is also key for these processes is not so intuitive, yet data supporting this idea began accumulating during the last century. Pathologists had observed dying cells in certain healthy tissues, and the term "apoptosis" was coined in 1972 to describe the phenomenon.[5] It was the discovery of the role cell death plays in the development of the nematode worm, however, that confirmed a positive function for apoptosis. The worm egg undergoes highly defined cell divisions that generate an adult comprised of a fixed total number of cells. Monitoring of cell lineages revealed that certain cells always undergo apoptosis at specific stages of development. Also, mutants were isolated in which apoptosis was suppressed; they ended up with more cells. These experiments led to the concept that there is a program of cell death. At specific developmental stages, the cell death pathway switches on in certain cells and a cascade of proteases is activated, leading to the lysis of these cells. In recent years, researchers have identified many genes underlying this built-in cell death program.

Cells that don't kill themselves for the greater good can go wrong in a thousand ways. In the mid–nineteenth century, Remak and Virchow argued that the cells of diseased tissues are derived from normal tissues. The implication was that malfunctioning cells beget disease. This argument has been of great medical importance, because it focused attention on changes in cellular (and ultimately molecular) behavior as critical factors for understanding disease.

This strategy is well illustrated by the study of cancer. Early pathologists recognized that cancer arises in abnormally developing tissues in which cell reproduction has become uncontrolled. By the 1970s, researchers had shown that genetic alterations, such as chromosomal rearrangements and the presence of certain viral genes, are responsible for cells becoming cancerous.[6] The identification in the 1970s of the *src* oncogene in the chicken Rous sarcoma virus, and later of its mammalian counterparts, was an important discovery. Many other oncogenes, which promote cell growth and division when activated, were subsequently discovered.

Tumor-suppressor genes, which normally restrain cell growth and division, also were identified. When these become inactivated in cells, those cells become cancerous. Other cancer-related genes participate in safety circuits, including ones that monitor gene damage and activate the cell death pathway if the genetic damage to individual cells is so great that it might induce cancer. The gene encoding p53, the most frequently mutated gene in human cancer, has become known as the "guardian of the genome," because it's required for proper surveillance of genomic damage. Cells lacking *p53* survive even if they have suffered extensive gene damage.[7] Cancer cells can also be defective in the cell death pathway; in these cells, apoptosis cannot be activated, so cells with genetic damage survive. If these surviving cells have damage to genes that control cell growth and division, they will selectively proliferate and dominate the tissue of origin. Eventually, they may undergo metastasis and spread to other parts of the body.

Round and Round She Goes: The Cell Cycle

Because cell reproduction is the basis for heredity, growth, and development, the cell needs to replicate and segregate all the genes so that the entire genome can properly pass on at each cell division. This is achieved by a series of events called the cell cycle.

The two major cell cycle events required for the replication and segregation of the genome are S-phase (for DNA synthesis) and M-phase (for mitosis).[8] During S-phase, the cell makes a copy of its chromosomal DNA. The process originally was identified by labeling studies and by making precise measurements of changes in cellular DNA content during the cell cycle.[9, 10] These experiments showed that DNA synthesis occurred only within a limited period early in the cycle. During M-phase, which occurs at the end of the cell cycle, the replicated chromosomes segregate into the two daughter cells formed at cell division. The periods between the birth of the cell and S-phase, and between the end of S-phase and M-phase, are respectively known as G_1 and G_2 (G stands for gap), leading to a four-phase cell cycle: G_1, S-phase, G_2, and M-phase.

There's an industrial park's worth of molecular machinery running the cell cycle. Consider DNA replication. Nothing revealed more about this process than the 1953 discovery of the structure of DNA.[11] This revelation solved the twin problems of how DNA could be precisely replicated and how it could encode information. DNA's sequence of bases encodes the information; the unwinding of the two-stranded, base-paired structure yields a pair of single-stranded templates for the synthesis of exactly complementary strands.

With DNA's structure in hand, the search for the machinery behind cellular replication became easier. The discovery of an enzyme, DNA polymerase, which could synthesize a DNA strand complementary to a preexisting DNA template, provided the first key.[12] Many other relevant enzymes were subsequently discovered, among them topoisomerases and helicases, which unwind DNA strands; primases and DNA polymerases, which synthesize new DNA strands; and ligases, which link strands together.[13] During a cell's S-phase,

1953
Francis Crick and James Watson publish the structure of DNA.

1956
Arthur Kornberg characterizes DNA polymerase, which can synthesize DNA.

1960s
Jacque Monod and François Jacob describe feedback control.

1970
Peter Duesberg and Peter Vogt discover the first oncogene, dubbed *src*, in a virus. The gene subsequently is implicated in many human cancers.

1971
Albert Knudson discovers the first tumor-suppressor gene.

Yoshio Masui identifies MPF (maturation promoting factor), which advances frog oocytes into M-phase.

1972

Andrew Wyllie and colleagues coin the term *apoptosis* for the built-in program of cell death, which subsequently is found to be important in development and disease.

1970s

J. Michael Bishop and Harold Varmus link cellular oncogenes with cancer.

Arnold Levine, Lionel Crawford, and David Lane discover *p53*, the gene most commonly mutated in human cancer.

1970s and '80s

Lee Hartwell develops yeast cell cycle mutants and the checkpoint concept; Paul Nurse defines cell cycle rate-limiting steps and discovers the cyclin-dependent kinase cdc2 in fission yeast and humans.

1983

Tim Hunt discovers cyclins.

1984

Marc Kirschner and Tom Mitchison describe the dynamic instability of microtubules.

these enzymes assemble on specialized regions of DNA, called "origins of replication," from which replication of the DNA extends bidirectionally. The two trajectories of replication move apart and form "bubbles" of DNA, which ultimately fuse to complete the process.

Like any complex machinery, the replication machinery requires control. Before replication can begin, for example, initiating factors have to bind to the origins of replication. These factors include origin recognition complexes, which act as "landing pads" for initiator proteins. These, in turn, load other proteins required for DNA replication. Normally, this process is activated only once in each cell cycle so only one S-phase takes place. Otherwise, cells might copy their chromosomes too many times.

Proper DNA replication is the first major component of the cell cycle's reproductive machinery. The second operates during M-phase, mitosis, first described in the nineteenth century.[4] Condensed replicated chromosomes, each consisting of two sister chromatids paired along their length, become visible at the beginning of mitosis. The chromosomes line up in the cell's middle and become associated with the mitotic spindle. The spindle extends between two structures, called centrosomes, located at opposite ends of the cell. The chromatids then move apart toward their respective sides of the cell, where they then segregate into two nuclei to be inherited by the newly forming daughter cells.

Central to this mechanism are tubulin polymers, which form the microtubules of the mitotic spindle. These microtubules exhibit dynamic instability—they undergo transitions between growth and shrinkage.[14] The centrosomes stabilize the microtubules and thereby the spindle. Other growing microtubules emanating from the centrosomes "explore" the cell's interior until their ends become stabilized by association with the kinetochore, a structure at the centromere found on each chromatid. That process links each chromatid to a centrosome.

Stable linkage can occur only if the kinetochores on the respective sister chromatids are linked to the cell's two centrosomes. When this happens, each half of each paired chromatid becomes oriented toward opposite ends of the cell. Cohesion between sister chromatids subsequently is lost, allowing the chromatids to separate toward opposite centrosomes. The act of separation unfolds by the combined actions of microtubular shrinkage and motor proteins that impel the microtubules to slide over each other. This process links the replication of DNA at the molecular level to the separation of chromosomes at the cellular level and ensures the precise replication and segregation of the genome.

Checks and Balances

To work properly, this molecular machinery requires regulation. Our understanding of the cell cycle's control systems has emerged from two concepts. The first one, the idea of checkpoints, began with the proposal that the cell cycle is a sequence of dependent steps, the initiation of later events occurring only after the successful completion of earlier ones.[8, 15] The notion of checkpoints was developed by studies with budding yeast.[16] This work revealed how the cell "checks" at particular "points" whether early events in its cycle have been completed properly. If not, the cell sends signals that block later events. For example, an incomplete S-phase leads to the sending of a signal that prevents mitosis. This spares the cell from undergoing a potentially lethal mitosis with only partially replicated DNA.

Other checkpoint controls monitor for DNA damage and block S-phase or mitosis until repairs are made. Researchers have found that damaging DNA or blocking its replication in yeast cells prevents mitosis and cell division. They also have found mutant yeast cells that allow cell division to proceed despite such problems. It turns out that these mutants express genes that act in checkpoint pathways, which monitor DNA damage and replication.[7] One example of such a gene in mammalian cells is *p53*.

The second major concept to deepen our understanding of cell cycle control is that of rate-limiting steps. This concept emerged from studies of fission yeast cells and frog oocytes. The discovery of yeast mutants that prematurely rush cells into mitosis led to a proposal that these mutants had altered versions of genes that regulate the timing of mitosis.[17] Complementary work with frogs identified protein factors that hasten frog oocytes into M-phase, suggesting that these factors were also rate-limiting controls for cell cycle progression.[18] The proteins encoded by the genes identified in yeast, and the protein factors identified in frogs, turned out to be the same—cyclin-dependent kinases, or CDKs.[19]

CDKs act as the "cell cycle engine," driving cells through S-phase and mitosis.[7] They are enzymes comprising a catalytic kinase subunit and a regulatory cyclin subunit.[20] The task of CDKs is to phosphorylate, and thereby control, other proteins required for the onset of S-phase and mitosis. CDK activity is itself regulated by such factors as inhibitors and the availability of cyclins. As cells proceed through a cycle, relevant CDK activities take place; different activities promote the cell through G_1, initiate S-phase, and initiate mitosis. For cells to exit mitosis and enter the G_1 phase of the next cycle, CDK activity must drop. This happens when the cyclin subunit of CDK degrades.

1986
Bob Horvitz proposes programmed cell death during development of the nematode worm.

1991
Andrew Hoyt and Andrew Murray discover spindle checkpoints.

1990s
Mechanisms of initiation of eukaryotic DNA replication begin to be elucidated.

Besides pacing the cell cycle, CDKs have a hand in checkpoint controls.[7] For example, the signaling pathway that blocks mitosis if DNA is damaged or if replication is incomplete probably operates in many cell types by preventing an increase in CDK activity. Once these flaws are rectified, CDK activity increases and mitosis can follow. CDK activity has yet another regulatory role: preventing a further S-phase in G_2 cells until mitosis is finished. Probable targets for this control are the initiator proteins. Blocking them from loading the proteins required for DNA replication would prevent an S-phase.

Not all checkpoint controls monitor for DNA damage or replication flaws. Genome stability also requires faithful chromosome transmission at mitosis; a checkpoint for this process stops mitotic progression if the spindle is not properly assembled or if the chromosomes are not properly attached to the spindle.[7] The likely role for this checkpoint control is to verify whether microtubules are securely attached to the kinetochores and, if not, to maintain the cohesion between sister chromatids so they cannot separate. This delays mitotic progression until all of the sister chromatids become both properly associated with the spindle and linked to the appropriate centrosomes.

This remarkable set of cell cycle and checkpoint controls helps ensure that cells proceed smoothly through their cycles. Without them, the genome could not be precisely replicated and segregated to provide each newly divided cell with a full complement of genes. The payoff is genomic stability.

Discovering Life's Chemistry [2, 21–24]

Combining these characteristics of cell reproduction with Darwin's theory of evolution by natural selection yields a framework for defining life. This view, first articulated by Hermann Muller in the 1920s, argues that living things have properties allowing them to undergo natural selection and thus to evolve. First, living organisms must be able to reproduce. Second, they must have a heredity system that can pass on the information defining the properties of the organism. Third, the heredity system must involve some variability, which also can be passed on. This variability provides the biological differences on which natural selection operates.

These properties of organisms are also characteristics of cells. Cells reproduce; they possess a heredity system based on genes; and replication errors or DNA damage elicits genetic changes that will be inherited during cell division. These properties allow cells to evolve, and because cell reproduction is the basis for all biological reproduction, it follows that these same cellular properties are the basis of the evolution of all living organisms. That holds even for exotic life forms that may harbor non–DNA-based heredity systems. Speculative possibilities include replicating clay particles as primitive life forms and organisms in other solar systems.[25]

Muller's definition of life is attractive, especially to geneticists. But it is not useful for explaining cellular functions, such as metabolism and growth. Making sense of these requires looking into the chemistry going on within cells.

During the second half of the nineteenth century, studies of fermentation showed that living cells promote specific chemical reactions. Humans had enjoyed the fermentation of crushed fruits and seeds for millennia, but it was Antoine Lavoisier who in the eighteenth century first recognized that fermentation was due to chemistry. Subsequent microscopic examination of ferments led Theodor Schwann and Charles Cagniard-Latour to propose in 1835 that the force driving fermentation was associated with a living organism, the yeast microbe. Louis Pasteur developed this idea by arguing that different microbes carried out specific fermentations, yielding different chemical substances. By 1858 he had concluded that fermentation was a physiological act that gives rise to multiple chemical products necessary for the life of the cell.

The next major insight about biological chemistry was the discovery that enzymes are behind it all. In 1860 Moritz Traube speculated that substances promoting fermentation reactions reside within cells and that their action is analogous to the soluble ferments known to exist outside cells. Within two years, Pierre Berthelot had provided experimental support for this speculation. From macerated yeast, he obtained a soluble ferment, called invertase, which could chemically degrade sucrose. About 30 years earlier, Jöns Berzelius had proposed the concept of catalysis, presciently providing the theoretical background necessary to appreciate Berthelot's experiments. Berzelius postulated the existence of catalytic substances, which behave like heat in promoting chemical transformations. He even suggested that thousands of different catalytic processes occur in animals and plants.

The stage was set for the key advance made in 1897 by the Buchner brothers, Eduard and Hans. They showed that a filtrate of a yeast extract contained a catalytic substance (zymase), which could promote in vitro the chemical reactions characteristic of fermentation. This discovery became the cornerstone of biochemistry. It led to the view that enzymes are protein catalysts that generate the compounds inside cells. Metabolism and cell growth, therefore, eventually were reinterpreted in the context of chemical reactions orchestrated by catalytic enzymes whose properties are specified by their associated genes.

Cells have to be innovative to allow such a myriad of chemical reactions to unfold within them even though many different conditions—including differing pHs and ionic environments—are required for the different enzymes to work well. Furthermore, for the cell's more complex chemical syntheses and molecular actions, several enzymes have to work in tandem to carry out a sequence of activities. For this, cells have to maintain a range of distinct microenvironments within different cellular compartments. This is a central principle underlying cellular organization.

To achieve this organization, the cell first has to maintain the general conditions necessary for life's chemistry; it must be insulated from the local environment. The cell does this with its outer lipid membrane, which is equipped with pumps and transporters that manage the movement of molecules into and out of the cell. Internal compartmentalization is the key to maintaining different microenvironments, and there are a number of mechanisms that generate these. First, enzymes have charged and precisely shaped surfaces, which establish a local environment conducive to specific chemical reactions. At a higher level of organization, enzymes link together to carry out sequential reactions. An example is the process of metabolic channeling. The product from one enzymatic reaction in a metabolic pathway becomes the starting material for the next enzyme in the pathway.

Similarly complex processes can be carried out by molecular machines consisting of the appropriate enzymes held in place within scaffolds made of protein and RNA. Such arrangements can undergo reconfigurations that promote sequential reactions. Examples are ribosomes, which carry out protein syntheses, and the complex of enzymes underlying DNA replication. Organelles such as the Golgi apparatus, the mitochondrion, and the nucleus embody yet a higher level of spatial organization. These are lipid-bound compartments, often with pumps and transporters, which maintain different microenvironments within the cell.

Spatial organization is a major cellular strategy for managing chemistry. Another is temporal organization. Examples here are DNA replication and mitosis. Both involve chromosomes but at different times during the cell cycle. DNA replication requires the chromosomes to be decondensed to allow access for enzymes; but mitosis requires the chromosomes to be condensed to foster their movement. The cell's solution is to separate the two processes temporally: DNA replication occurs early in the cell cycle and mitosis occurs later.

For the past century, ever more of life's functions have been interpreted in terms of cellular chemistry. As Jacques Loeb argued in his 1912 essay collection, *The Mechanistic Conception of Life,* the cell is like a chemical machine.

Enforcing Discipline[24]

It takes regulation and coordination to run all of the chemistry going on inside cells. Analyses of enzyme and gene activity, as well as the rise of cybernetics during the mid-twentieth century, put a spotlight on the concept of feedback control as a strategy for such regulation. The general idea was that the output of a process has regulatory consequence on the process itself. The simplest examples are quantitative, in which products generated by a metabolic pathway or by gene activity feed back to an earlier stage of the process, thereby slowing it down or speeding it up. The idea of feedback can also be applied in quality-control contexts. Consider the proofreading mechanism associated with the enzyme complexes that replicate DNA.[13] This complex corrects errors of base-pair insertion that occur during the syntheses.

Such regulatory activities are closely associated physically with the process being regulated. That's not so when there is a need to coordinate activities that unfold at different cellular locations or at different times in a cell's history. Here, regulation requires a communication system that extends over substantial distances and times. One strategy is to send molecular signals from the cell's outer membrane to the nucleus, a process that affects gene expression. Other signals emerging from cellular checkpoint controls, which register problems during S-phase, serve as "reminders" to delay M-phase later in the cell cycle.

This kind of higher-level organization distinguishes the chemistry of life from that of nonliving systems. It also highlights two features of cells: the ability to assume spatial organization and the ability to process information through signaling networks. Direct molecular interactions can lead to low-level spatial organization, such as the construction of cellular machines like ribosomes. But much of the cell's larger spatial organization requires more. How this extended organization is achieved remains unknown. However, chemical physicists and theoretical biologists who study the spatiotemporal organization of complex chemical reaction systems, such as the Belousov-Zhabotinsky reaction, have been finding clues. The network of chemical reactions underlying this remarkable system leads to changes in the concentrations of intermediates that oscillate in time and that generate standing waves. This results in patterns of chemicals within the reaction vessel. A primary goal for biologists is to explain how chemical reactions bring about analogous organization on cellular scales.

One interesting approach is to focus on the generation of dynamic chemical structures that require energy input for their maintenance and that involve fluxes of molecular components through the structures. Changing the kinetics of these processes can lead to structures of different sizes and shapes, features that presumably could be regulated in response to cellular needs. These ideas are more than metaphors for thinking about cellular organization; scientists are applying them to help explain the generation of structure and behavior, such as mitotic spindles[26] and cell cycle regulation.[27]

The other feature of cells pertinent to higher-level organization is the processing of information through signaling networks. These are not always linear pathways; they can be complex, with a variety of inputs and outputs, with bifurcation and amplification steps, and with some components acting at more than one step in a pathway and with others acting in parallel. The overall significance of such complex networks for signaling behavior is yet to be fully appreciated. One intriguing issue is the potential for information to be carried within the dynamic aspects of signaling, as is seen in the Morse code, which encodes messages in temporal patterns. Another is the role of threshold levels, in which high-intensity and low-intensity signals each send different messages. Modelers of interconnected networks have shown that ordered signaling patterns can emerge from relatively simple wiring diagrams and rules of operation.[28] Exciting discoveries are in store for those who investigate how signaling pathways operate in the real space and real time of the cell.

The cell is the simplest unit to exhibit life's functions. We now have both the molecular tools and the conceptual frameworks to undertake a concerted program to understand how cells operate. The genome projects will anchor that foundation by identifying all the genes required for a cell to function, yet researchers will still have to work out how the relevant gene products act and interact to generate cellular organization. Particularly worthwhile will be investigations into those physiochemical properties that allow order to be generated on the extended scales within the cell, and studying the ways in which complex information networks operate. There are many different ways in which all the genes in the cell could function in concert, but only certain regulatory and organizational states are likely to be effective and stable, and therefore compatible with life. Uncovering the rules that govern these states and the reasons why they exist are worthy research goals. The coming years will be exciting ones during which new ideas and theories will help us fully understand cells and thereby life itself.

References

1. H. Harris, *The Birth of the Cell* (Yale University Press, 1999).
2. W. Coleman, *Biology in the Nineteenth Century. Problems of Form, Function, and Transformation* (Cambridge University Press, London and New York, 1977).
3. J. R. Baker, *The Cell Theory. A Restatement, History, and Critique* (Garland Publishing Inc., New York, 1988).
4. E. B. Wilson, *The Cell in Development and Heredity* (Garland Publishing Inc., New York, 1987).
5. D. L. Vaux and S. J. Korsmeyer, *Cell* 96, 245 (1999).
6. H. Varmus and R. E. Weinberg, *Genes and the Biology of Cancer* (Scientific American Library, New York, 1993).
7. A. Murray, T. Hunt, *The Cell Cycle, an Introduction* (Oxford University Press, Oxford, U.K., 1993).
8. J. M. Mitchison, *The Biology of the Cell Cycle* (Cambridge University Press, London and New York, 1971).
9. H. Swift, *Proc. Natl. Acad. Sci. U.S.A.* 36, 643 (1950).
10. A. Howard and S. Pelc, *Heredity* 6, 261 (1953).
11. J. D. Watson and F. H. Crick, *Nature* 171, 737 (1953).
12. A. Kornberg, I. Lehman, M. Bessman, E. Simms, *Biochim. Biophys. Acta* 21, 197 (1956).
13. A. Kornberg and T. A. Baker, DNA *Replication* (Freeman, New York, 1992).
14. T. Mitchison and M. Kirschner, *Nature* 312, 237 (1984).
15. L. Hartwell, *Bacteriol. Rev.* 38, 164 (1974).
16. L. H. Hartwell and T. A. Weinert, *Science* 246, 629 (1989).
17. P. Nurse, *Nature* 256, 547 (1975).
18. Y. Masui and C. Markert, *J. Exp. Zool.* 177, 129 (1971).
19. P. Nurse, *Nature* 344, 503 (1990).
20. T. Evans, E. Rosenthal, J. Youngbloom, D. Distel, T. Hunt, *Cell* 33, 389 (1983).
21. D. Dressler and P. Huntington, *Discovering Enzymes* (HPHLP/Scientific American Library, New York, 1991).
22. G. Allen, *Life Science in the Twentieth Century* (Cambridge University Press, London and New York, 1978).
23. B. Alberts et al., *Molecular Biology of the Cell* (Garland Publishing, New York, Ed. 3, 1994).
24. B. Lewin, *Genes VII* (Oxford University Press, Oxford, U.K., 1999).
25. J. Maynard Smith and E. Szathmary, *The Major Transitions in Evolution* (W. H. Freeman, Oxford, U.K., 1995).
26. M. Kirschner, J. Gerhart, T. Mitchison, *Cell* 100, 79 (2000).
27. B. Novak and J. J. Tyson, *J. Theor. Biol.* 173, 283 (1995).
28. S. Kauffman, *The Origins of Order* (Oxford University Press, Oxford, U.K., 1993).

29. Thanks to Jacky Hayles, Teresa Niccoli, and John Tooze for their critical reading of this chapter.

10

The Ascent of Atmospheric Sciences

PAUL J. CRUTZEN AND VEERABHADRAN RAMANATHAN

Ever since early generations of scientists began teasing apart the unseen components of the air we breathe, atmospheric science has been on the rise. In the past few decades, as phrases such as "stratospheric ozone depletion" and "greenhouse warming" became part of everyday discourse, atmospheric scientists too have become more publicly visible and important. In this essay, Paul J. Crutzen and Veerabhadran Ramanathan chart some of the pathways by which this field evolved from its beginnings in the seventeenth century into a multidisciplinary enterprise that routinely generates globally consequential knowledge.

PAUL J. CRUTZEN is Professor Emeritus of Atmospheric Science at the Max Planck Institute for Chemistry in Mainz, Germany. He was a co-recipient of the 1995 Nobel Prize in chemistry for his contributions to the understanding of stratospheric ozone production and destruction. VEERABHADRAN RAMANATHAN is Professor of Atmospheric Sciences at the Scripps Institute of Oceanography, which is part of the University of California in San Diego. He also is director of the University's Center for Atmospheric Sciences. His research includes investigations into climate change and the consequences of human activity on the environment.

TIMELINE OF ATMOSPHERIC SCIENCES

C. 340 B.C.
Aristotle's *Meteorologica* is published; its theories remain unchallenged for nearly 2,000 years.

1590s
Galileo Galilei invents the thermoscope, a precursor of thermometers.

1643
Using a mercury barometer, Evangelista Torricelli performs experiments on atmospheric pressure.

1686
Edmond Halley shows that low latitudes receive more solar radiation than higher ones and proposes that this gradient provides forcing for the atmosphere's general circulation.

1714
Gabriel Daniel Fahrenheit develops the thermometer and a temperature scale.

Atmospheric science matters to everyone every day. It has been infiltrating public awareness lately for compelling reasons: the Antarctic ozone hole, global warming, and El Niño, a combined atmosphere-ocean phenomenon that causes severe weather. The first two are side effects of the industrial revolution, and El Niño is nature's warning against taking good weather for granted. Atmospheric science has become a multidisciplinary, high-tech activity rife with new and sophisticated instrumentation, computers, information technology, and measurement platforms, including satellites and aircraft.

Air chemistry,[1–4] including the prediction and subsequent verification that chlorofluorocarbons (CFCs) destroy ozone, has recently moved to the field's forefront. Revelations about the susceptibility of people and ecosystems to both natural and humanly forced atmospheric changes also have been central. On a more purely scientific level, efforts to understand and predict the complex phenomenon of weather led to the discovery of chaos theory by meteorologist Edward Lorenz. Now, chaos theory helps physicists, chemists, biologists, economists, and many others striving to understand complex phenomena.

A comprehensive story about the ascent of atmospheric sciences over the past few centuries would fill library shelves. In these few pages, we first chronicle only a few developments in chemistry and meteorology up to the early 1970s, before the possibility of human influence beyond the local scale became actualized. To illustrate how humanity's hand has grown to have global effects, we then zero in on two contemporary issues: atmospheric ozone and global warming.

Secret Air

Even by the mid–seventeenth century, the properties of air had received remarkably little thought. Robert Boyle knew as much at the time as anybody when he opined that the atmosphere consists not of the "subtle matter," or ether, that his contemporaries presumed to fill the universe, but primarily of "exhalations of the terraqueous globe," that is, emanations from volcanoes, decaying vegetation, and animals. By the end of the nineteenth century, however, research into the nature of air already had proven invaluable. About a century ago, William Ramsay assessed the situation this way: "To tell the story of the development of men's ideas regarding the nature of atmospheric air is in great part to write a history of chemistry and physics."

Isolation of air's actual constituents began in the eighteenth century. Joseph Black identified carbon dioxide in the 1750s; Daniel Rutherford isolated nitrogen in 1772. A few years later, Carl Scheele and Joseph Priestley independently discovered oxygen. In 1781, Henry Cavendish established the composition of air to be 79.16 percent nitrogen and 20.84 percent oxygen regardless of location and meteorological conditions. By 1894 Lord Rayleigh and Ramsay had discovered argon in air. Thousands of chemical compounds have now been detected in the air, many at concentrations even less than 10^{-12} in molar mixing ratios (less than 1 part per trillion).

In 1839,[3] Christian Schönbein literally smelled a previously unidentified air component that would prove central to the scientific and public interest. During electrolysis experiments with water, he noticed a sharp odor. He ascribed it to a compound, which he called "ozone," after a Greek word meaning "ill smelling." In 1864 Jacques L. Soret recognized that ozone was the "dioxide of the O atom." In 1878, Alfred Cornu noticed the absence in sunlight of wavelengths shorter than 310 nanometers and postulated the presence of an absorbing gas in the atmosphere. Two years later, Walter N. Hartley measured the spectral properties of ozone and concluded that it was Cornu's absorber. The absorption of this ultraviolet (UV) radiation is essential for the survival of many of the planet's life-forms.

In 1902, Léon-Philippe Teisserenc de Bort discerned another important clue about ozone. Making measurements aboard a balloon, he found that at altitudes above 8 to 10 kilometers temperature did not further decline with altitude as it did at lower altitudes. He called this regime of more constant temperature, which is maintained by ozone's absorption of solar UV radiation, the stratosphere; the region below he dubbed the troposphere.

In the 1920s, Gordon M. B. Dobson measured the total amount of ozone in vertical columns of atmosphere. He found maximum amounts at higher latitudes and during early spring. These findings showed that stratospheric circulation plays a major role, implying upward transport from the troposphere to the stratosphere in the tropics with the return flow at middle and higher latitudes, the so-called Brewer-Dobson (after Alan Brewer) circulation. The theoretical framework for this circulation was developed over the next 50 years. A unified picture of the dynamical processes responsible for stratosphere-troposphere exchange of ozone and other gases was presented in 1995 by James Holton, Peter Haynes, Michael McIntyre, and their colleagues. The upward transport through the cold region of the tropical troposphere causes the stratosphere in general to be dry and cloud-free.

In 1930, the field of atmospheric photochemistry[3–5] began when Sydney

1735
George Hadley introduces the importance of Earth's rotation in the features of major winds, including the easterly trades and the midlatitude westerlies; the tropical Hadley circulation is named after him.

1750s
Joseph Black identifies CO_2 in atmospheric air.

1770s
Daniel Rutherford identifies nitrogen in air. Carl Scheele and Joseph Priestley independently discover oxygen.

1781
Henry Cavendish measures the composition of air to be 79.16 percent nitrogen and 20.84 percent oxygen.

1839
Christian Schönbein discovers ozone in the laboratory.

1859

John Tyndall suggests that water vapor, CO_2, and any other radiatively active ingredient of the atmosphere could produce "all of the mutations of climate which the researches of geologists reveal."

1878

Alfred Cornu measures the spectrum of solar radiation reaching Earth's surface and suggests that the absence of short-wave ultraviolet radiation is due to an atmospheric absorber. Two years later, Walter Hartley concludes that Cornu's absorber is ozone in the upper atmosphere.

1896

Svante Arrhenius publishes a climate model demonstrating the sensitivity of surface temperature to atmospheric CO_2 levels.

1919–25

Jacob Bjerknes, his father Vilhelm, and Halvor Solberg develop polar front theory for describing local meteorology.

Chapman proposed that ozone is produced photochemically from oxygen. Solar UV radiation with wavelengths shorter than 240 nanometers breaks O_2 molecules into two oxygen atoms, each of which then combines with another O_2 to yield ozone, O_3. The reverse process occurs when oxygen atoms from the photolysis of ozone react with O_3 molecules, resulting in O_2. David Bates and Marcel Nicolet extended this scheme in 1950 by showing that hydroxyl-based radicals derived from the photolysis of water molecules can catalytically convert ozone and atomic oxygen into O_2.

Air chemistry is just one—in fact one of the youngest—of the atmospheric sciences. The invention of the thermometer by Galileo Galilei and by others in the 1590s, and the 1643 invention of the barometer by Galileo's disciple Evangelista Torricelli, marked the start of experimental meteorology.[6,7] By the end of the seventeenth century, Edmond Halley had recognized the role of solar heating in the trade winds. His solar heat budget estimates revealed that the tropics receive much more solar radiation than the high latitudes. By the mid–eighteenth century, George Hadley adopted Halley's idea and concluded that equatorial air must respond to the more intense solar heating first by rising and then by drifting poleward before returning to lower levels at the equator. Hadley also developed the correct explanation for the trade winds by deducing that Earth's rotation deflects the equator-directed surface flow toward the west relative to Earth.

During the nineteenth century, observational technology took another leap, this time in the form of instrumented balloons. Significant advances in observation tools took place during World War II, including weather radar and "radiosondes," or instrumented balloons bearing radio transmitters. The new observations revealed large, longitudinally asymmetric motions in the atmosphere, such as extratropical cyclones, and the midlatitude jet stream. Armed with the new data, Jacob Bjerknes, Carl-Gustaf Rossby, and Jule Charney transformed the science of atmospheric motion into its modern form between 1920 and 1970.[8,9]

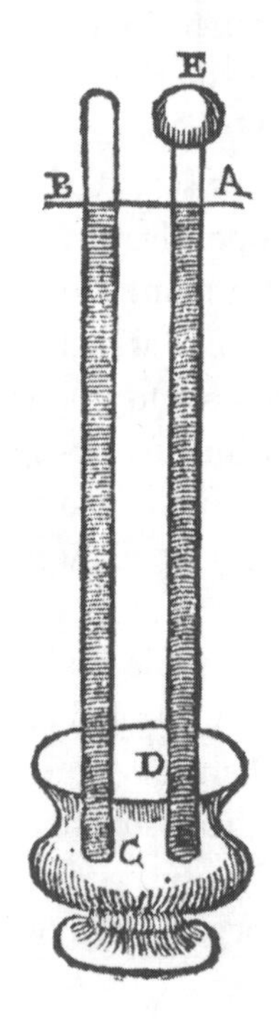

Under pressure. Using his newly invented barometer, Evangelista Torricelli began probing atmospheric pressure in the mid-seventeenth century. *E. Banfield, Barometers (Baros Books, Wiltshire, UK, 1995).*

Basically, the equator-to-pole gradient of radiative heating that Halley discerned leads to gradients in temperature, potential energy, and zonal air flows, such as the upper troposphere jet stream. The jet is unstable to small perturbations, and the resulting wavelike eddy motions grow and concentrate temperature gradients to form narrow warm and cold fronts, which generate much of local weather. The poleward transfer of heat in the extratropics is accomplished by these eddies, without which the polar regions would be more than 50°C colder and the tropics would be much hotter.

With a theoretical foundation for atmospheric circulation coming into place, another question arose: How is surface temperature regulated?[10] Jean-Baptiste Fourier suggested in 1824 that the atmosphere behaves like the glass cover of a box exposed to the sun by allowing sunlight to penetrate to Earth's surface and then retaining much of the "obscure radiation" (wavelengths greater than about 4 µm that emanate from Earth's surface). By the mid–nineteenth century, John Tyndall had demonstrated that the key process retaining this longwave radiation is its selective infrared (IR) absorption by atmospheric H_2O and CO_2.

Constructing such greenhouse models of the atmosphere required accurate measurements of solar and IR transmission. That's where Samuel Langley's observations of lunar and solar spectra from 1885 to 1890 came in. Those data, and the identification of numerous H_2O and CO_2 absorption bands in spectroscopic measurements, set the stage for the famous study in 1896 by the Swedish chemist Svante Arrhenius. He developed a detailed model for the surface-atmosphere radiation budget and used it to reveal the large sensitivity of surface temperature to increases in atmospheric CO_2. Arrhenius included the positive feedback due to water vapor (the dominant greenhouse gas), whose atmospheric loading increases with temperature because of the increase in saturation vapor pressure with temperature. As a result, a water-vapor–based greenhouse effect increases with a warming.

In 1967, Syukuro Manabe and Richard Wetherald reformulated the greenhouse theory. They demonstrated that surface temperature is not determined solely by the energy balance at the surface (as assumed by Arrhenius and others), but also by the energy balance of the surface-troposphere-stratosphere system. The underlying concept is that the surface and the troposphere are so strongly coupled by convective heat and moisture transport that the relevant forcing governing surface warming is the net radiative perturbation at the top of the troposphere. This concept of a radiative-convective equilibrium, which is used extensively in astrophysics, was originally proposed in 1862 by Lord Kelvin to explain the temperature decrease in the troposphere.

1919–37

Vilhelm and Jacob Bjerknes and others identify extratropical cyclones as waves on the polar front. Together with later planetary wave theories by Carl-Gustaf Rossby and Jule Charney, these atmospheric motions provide the modern-day dynamical foundation for understanding general circulation and weather forecasting.

1920

Milutin Milankovitch publishes his theory of ice ages based on variations in Earth's orbit.

1924–28

Gordon Dobson documents the latitudinal and seasonal variation of ozone.

1928

Gilbert Walker describes the "Southern Oscillation," the seesaw pattern of atmospheric pressure on the eastern and western sides of the Pacific Ocean.

1930
Sydney Chapman proposes the first photochemical theory for upper atmospheric ozone production.

1941
Radar is used for weather tracking.

1949–56
Work by Alan Brewer and Dobson leads to a model for stratospheric general circulation.

1950
David Bates and Marcel Nicolet propose that H, HO, and HO_2 play a catalytic role in the atmosphere.
Charney and John von Neumann demonstrate the science of numerical weather forecasting with the ENIAC computer.

1956
Norman Phillips completes the first successful numerical simulation of atmospheric circulation.

Another big part of the atmospheric and climate equations is clouds.[10] For decades, models have predicted that clouds have a net cooling effect. Data from the NASA Earth Radiation Budget Satellite Experiment confirmed those predictions in 1989. The data reveal that solar radiation reflected by clouds exceeds their greenhouse effect by a significant 15 to 20 $W.m^{-2}$—a cooling effect about five times larger than the warming effect from a doubling of CO_2. Small changes in cloudiness, therefore, can have large but uncertain feedback effects on climate. A comprehensive study by Robert Cess and colleagues in 1990 revealed that cloud feedback simulations using so-called general circulation models (GCMs) yield varying results. Until there is a theory or a model of the hydrological cycle that can accommodate the scales of all cloud systems—from tens of meters (cumulus) to more than 1000 kilometers (tropical cirrus)—rigorous climate prediction will remain out of reach.

There are other obstacles that have been keeping a detailed quantitative understanding of climate and the general circulation at bay. After all, these phenomena depend on numerous complex physical processes, among them radiative transfer; land surface processes; and the formation of clouds, ice, and snow cover. Computers began coming to the rescue in the mid-1950s, and thanks to their ever-increasing power, today's GCMs are starting to account for many of these processes, albeit not reliably yet. In the 1950s, weather prediction drove the development of GCMs (by Jule Charney, John von Neumann, and Norman Phillips), but once such models were available they were put to wider use for such tasks as simulating general atmospheric circulation, climate, and global warming.

Growing Impacts of Human Activities: Ozone and CFCs [4, 5]

Air pollution is as ancient as the harnessing of fire for cooking, heating, smelting metal, and clearing land. Diminishing wood resources in some European locations within the past millennium spurred coal use and with it a rise in sulfur-based air pollution, which has led to one of the most publicly recognized forms of pollution—acid rain.

Coal use increased 500-fold during the industrial period. The dearth of control and regulation enabled urban air pollution to worsen into a major problem, culminating in the killer smog of December 5–9, 1952 in London: An estimated 4,000 people died because of it. Since then, legislation has been passed to prevent a repeat. Still, with the continuing growth of coal and oil use in North America and Europe, and the construction of high chimneys to alleviate local pollution, the problem of acid rain has grown to international dimensions.

Based on data from a network of sites in Europe, which was established in the 1950s by the International Meteorological Institute in Stockholm, Svante Odén showed that highly acidic precipitation had spread to much of northern and Western Europe by 1968. Ecological consequences, such as forest decline and fish deaths, were most pronounced in the Scandinavian countries. The problem gained international attention in 1972 with the presentation of a Swedish study at the first U.N. Conference on the Human Environment in Stockholm. This study made a strong case for reducing sulfur emissions. Since then, Europe and North America have made considerable progress, but the use of sulfur-rich coals in parts of Asia is leading to a repeat of regrettable history.

In the 1940s and 1950s, the Los Angeles Basin began suffering from a new kind of air pollution, especially during sunny summer days. It caused poor visibility, eye irritation, and crop damage. Arie Haagen-Smit showed that this type of pollution could be simulated by irradiating mixtures of air and car exhaust with sunlight. "Photochemical smog" has now become commonplace around the world.

By 1970, the chemistry of ozone production near the surface (often called "bad" ozone, compared to the "good" ozone in the stratosphere that absorbs harmful solar UV radiation) had been identified. It begins with the attack by hydroxyl radicals (HO) on hydrocarbons or carbon monoxide, producing peroxy-radicals. These radicals donate O atoms to O_2 to form ozone, whereas NO and NO_2 are catalysts for ozone formation. The hydroxyl radicals themselves are produced when solar UV radiation infiltrates mixtures of ozone and water vapor in air.

Hiram Levy pointed out in 1972 that these radicals play a major role in atmospheric chemistry. Despite their rarity (the average molar mixing ratio of HO in air is 4×10^{-14}), they are known as the "detergent of the atmosphere." The lifetimes of most gases—whether derived from natural processes or human activities—are determined by their reactivity with HO and can vary from hours to years. Some gases, such as CFCs and nitrous oxide, do not react with HO. That's how they end up being so influential in stratospheric ozone chemistry.

Tropospheric ozone (about 10 percent of total atmospheric ozone) is central to atmospheric cleaning because of its role in creating HO. Contrary to the widespread belief that tropospheric ozone originates in the stratosphere,

1959

The *Explorer VI* satellite provides TV imagery of cloud cover. Verner Suomi uses *Explorer VII* to estimate the global radiation heat budget of the Earth-atmosphere system.

1963

Meteorologist Edward Lorenz derives a system of three equations for demonstrating the limits to the predictability of weather, which ultimately leads to the development of chaos theory.

1967

Syukuro Manabe and Richard Wetherald develop the one-dimensional radiative-convective model, including clouds, water vapor, CO_2, and ozone, and show that a doubling of atmospheric CO_2 can warm the planet by about three degrees. In 1975, these authors initiate GCM studies of climate change.

1969

Jacob Bjerknes links the Southern Oscillation and El Niño.

1970–74

Paul Crutzen and Harold Johnston identify an ozone-destroying catalytic cycle involving oxides of nitrogen (NO_x). Richard Stolarski and Ralph Cicerone propose another such cycle involving chlorine, and Mario Molina and F. Sherwood Rowland link this cycle to ozone depletion by chlorofluorocarbon (CFC) emissions.

1972

Hiram Levy postulates that reactions with hydroxyl radicals make up the main tropospheric sink for almost all gases emitted into the atmosphere.

1985

The British Antarctic Survey reports a systematic, springtime reduction of ozone over Halley Bay leading to a 40 percent drop between 1956 and 1985.

1987

Many countries sign the Montreal Protocol on Substances that Deplete the Ozone Layer, in which signatories agree to controls on the production and use of ozone-destroying materials.

one of us, Paul J. Crutzen, proposed in 1972 that the global troposphere itself is an important source of its own ozone and that human activity plays a major role.

The argument was based on tropospheric NO_x ($NO + NO_2$), much of which is produced by the burning of fossil fuel and tropical biomass. Depending on NO_x concentrations, large amounts of ozone can be produced or destroyed during the oxidation of ever-present carbon monoxide (CO) and methane (CH_4), levels of which have been rising. Together, these conditions have led to a substantial increase in tropospheric ozone, especially in the Northern Hemisphere. The tropics and the Southern Hemisphere also are influenced by human activity through extensive burning of biomass.

Human activity influences stratospheric ozone as well. In 1970, Crutzen and Harold Johnston called attention to the catalytic role of NO_x in controlling levels of stratospheric ozone and to the possibility of ozone destruction by NO-emitting supersonic aircraft. The rate-limiting reaction is O + NO_2ÆNO + O_2. The oxygen atoms come from the photolysis of O_3, and the NO_2 comes from the reaction NO + O_3ÆNO_2 + O_2. The net result is 2 O_3Æ3 O_2, or, in other words, ozone destruction.

Clearly, the role of NO_x species as catalysts in ozone chemistry is complex. We now know that above about 25 kilometers, NO_x catalyzes ozone destruction, but that the opposite holds at lower altitudes. The biosphere also is involved in the natural control of stratospheric ozone. That's because stratospheric NO forms by the oxidation of N_2O, which is produced largely by microbes in soils and waters. This is just one of many ways in which the biosphere is involved in atmospheric chemistry and climate, a complex dynamic most fully developed by James Lovelock in his much-debated Gaia hypothesis, which invokes self-regulating relations between atmospheric chemical composition, the biosphere, and climate.

As it turned out, large supersonic fleets were never built, but the research that emerged because of that possibility greatly improved knowledge about the stratosphere's chemistry and dynamics. This was important because in the meantime an enormous time bomb already had been ticking for several decades.

After their introduction in the 1930s as better and safer refrigerants, emissions of the CFCs $CFCl_3$ and CF_2Cl_2 began increasing. In 1972, Lovelock made the first worldwide CFC measurements on a research ship between England and Antarctica using his powerful electron capture technique, allowing measurements of gases with molar mixing ratios in the 10^{-12} range and below. He concluded from his data that $CFCl_3$ was accumulating in the atmosphere, but without consequences for the environment. Two years later, however, Mario Molina and F. Sherwood Rowland pointed out that when CFC gases break down in the stratosphere, they produce Cl and ClO. These radicals then attack ozone catalytically much as do NO and NO_2, but even more effectively.

With unabated CFC emissions, calculations suggested that ozone depletion would peak at altitudes near 40 kilometers, causing local losses up to some 30% to 40% after several decades. By the late 1970s, several nations, including the United States, Canada, Norway, and Sweden, had responded to this alarming finding by forbidding the use of CFC gases as propellants in spray cans. Their use for other purposes continued, however.

One reason for this modest response was that the Molina-Rowland mechanism was thought to result in ozone depletion largely above 30 kilometers. Most ozone is located at lower altitudes, however. And there, formation of the rather unreactive molecules, $ClONO_2$ and HCl, protects ozone from otherwise much greater destruction.

This sense of complacency rapidly evaporated in 1985 when Joe Farman of the British Antarctic Survey and colleagues published a most surprising set of measurements. Their data revealed 40 percent depletion of total ozone during austral spring compared to levels measured since 1956 at the Halley Bay Station in Antarctica. These depletions were much greater than predicted, and they occurred in a geographic region where such depletions were unexpected.

Subsequent measurement campaigns in the Antarctic—headed by David Hofmann, Susan Solomon, and James Anderson—revealed that the greatest ozone depletions occur at altitudes between 12 and 22 kilometers, exactly the range of naturally maximum ozone concentrations. Also, ClO and OClO radicals were found to be far more prevalent than predicted, suggesting that new chemistry was afoot. Solomon, Rowland, and their colleagues proposed that under the cold conditions prevailing in the ozone hole region, HCl and $ClONO_2$ molecules react on ice particles to yield Cl_2, which in sunlight dissociates rapidly into two ozone-destroying Cl atoms.

Through laboratory simulations Luisa and Mario Molina discovered yet new ozone-depleting catalysis, this one involving the reaction of ClO with another ClO as the rate-limiting step. This chemistry is especially important at altitudes below about 25 kilometers. Because annual increases in stratospheric chlorine were exceeding 5 percent, ozone depletion was increasing by twice that much. The Molinas' study and many others show the great importance of laboratory simulations of chemical reactions of atmospheric interest.[11, 12]

Further research showed that the chlorine activation reactions could occur on solid or supercooled liquid particles containing mixtures of water and

1989

The Earth Radiation Budget Experiment demonstrates that clouds reflect significantly more solar energy than the long-wave radiant heat energy they retain, thus exerting a large cooling effect on the planet.

1990s

Researchers establish the importance of anthropogenic aerosols in potentially offsetting the greenhouse effect by reflecting solar radiation to space.

The global warming trend that began in the 1970s continues, and average surface temperatures reach record values compared with the previous 100 years.

1995

An international team of more than 1000 scientists convenes under the charter of the Intergovernmental Panel on Climate Change and declares:

"The balance of evidence suggests a discernible human influence on global climate."

1996

The production of CFCs and some other ozone-depleting chemicals stops in the industrial world.

1997

The Kyoto Protocol calls for legally binding greenhouse gas emission targets for all developed countries.

sulfuric and nitric acids. Such particles can form at temperatures about 10°C higher than the freezing point of water ice from sulfuric acid–containing aerosol, whose ubiquitous presence and chemical composition in the stratosphere had been shown by Christian Junge in 1961. They also can be players in the Arctic. Depending on temperatures, ozone losses of up to 30 percent have occurred there in some years during the past decade in late winter to early spring. Total ozone loss over the Antarctic has been reaching the yet more dramatic levels of 50 to 70 percent each year during spring. After the discovery of the ozone hole, international regulations to limit CFC production picked up speed, resulting in a total phase out of production in the industrial world since 1996.

It is hard to tell if stratospheric ozone destruction or global warming is more sinister in the public mind. The realization that a doubling of CO_2 would warm the globe significantly inspired Arrhenius to resurrect Tyndall's suggestion that history's glacial epochs may be due to large reductions in atmospheric CO_2. Another impetus for investigating the connection between CO_2 and surface warming was Guy S. Callendar's conclusion in the late 1930s that fossil fuel combustion had been increasing atmospheric CO_2. In 1958, Charles D. Keeling began continuous measurements of CO_2 at the Mauna Loa Observatory and demonstrated beyond any doubt that CO_2 is increasing at a rate of about 1.5 parts per million (ppm) per year due to anthropogenic activities.

Then came Manabe's and Wetherald's 1975 GCM study, which showed that doubling CO_2 can warm the globe by up to 3 K. Nearly simultaneously, the global warming problem took off in another attention-getting direction. One reason was the identification by one of us, Veerabhadran Ramanathan, that CFCs have a direct greenhouse effect and that a molecule of $CFCl_3$ or CF_2Cl_2 was about 10,000 times or more effective than a molecule of CO_2 at enhancing the greenhouse effect.

In 1976, CH_4 and N_2O joined the list of greenhouse gases. A few years later, tropospheric ozone also joined the list, which has grown to include several tens of greenhouse gases. These developments culminated in an international study in 1986[13] sponsored by the World Meteorological Organization. The study concluded that the non-CO_2 gases significantly added to the warming by CO_2. The global warming problem suddenly had become more urgent.

The urgency was further accentuated by the fact that the surface warming trend, which began in the 1970s and continued unabated into the 1990s, had reached the point where the global mean surface temperature was higher than ever recorded during the twentieth century. James Hansen undertook GCM simulations that included observed trends in CO_2 and other trace gases. The simulated temperature trends supported scenarios in which the human input of such gases indeed was the cause of surface warming, a finding that placed global warming higher on the political agenda. The problem's global nature demanded international reviews. So in the late 1980s, the Intergovernmental Panel on Climate Change (IPCC) was established to make a comprehensive assessment.[14] It concluded that "the balance of evidence suggests a discernible human influence on global climate."

That point coincides with the finding that global warming is also influenced by stratospheric ozone depletion. As shown by Jeffrey Kiehl and Venkatachalam Ramaswamy, for example, the observed lower stratospheric cooling during the 1980s and 1990s is explained by the radiative response to ozone destruction, a process carrying implications for tropospheric climate as well.

An Uncertain Future and the Need for Synthesis

We start the new century with formidable environmental challenges that affect the welfare of all humans. On the potentially positive side, stratospheric ozone could recover and the ozone hole could disappear by midcentury, although we cannot rule out the possibility of it becoming worse before it gets better due to cooling of the stratosphere by increasing CO_2. On the negative side, by the early twenty-second century, atmospheric CO_2 is expected to double from its concentration of 280 ppm[14] prior to inputs from human industry.

Unfortunately, the negative possibility seems a good bet. After all, nearly 75 percent of the world's population by then, which is projected to approach 10 billion people, will be striving to match Western standards of living. Those efforts will likely entail enormous additions of atmospheric pollutants, land surface alteration, and other environmental stresses. The net radiant heat added to the planet (since 1850) by greenhouse gases is likely to amount to at least 4 $W.m^{-2}$. According to our best understanding of the system, that could warm the planet by about 1.5 to 3 K, depending on competing effects from aerosols. The atmosphere and the planet, it would seem, are headed toward uncharted territory.

One of the major surprises could very well be the response of El Niño[15] to such a warming. Warren Washington and Gerald Meehl have proposed that global warming can lead to an increase in the frequency of El Niños through feedback mechanisms involving clouds and ocean circulation in the Pacific. It may well be argued that the environmental expansion of human activity has jolted Earth into a new geological era, the "Anthropocene."

Temptations to engineer the atmosphere to mitigate the negative effects of human activities could become enormous. Might it be possible to combat global warming by burying CO_2 in the deep ocean, or by supplying iron to parts of the ocean to increase photosynthetic and oceanic CO_2 uptake, or by adding light-scattering particles to the atmosphere to scatter sunlight back into space? How about adding propane to counter stratospheric ozone loss? The practicality, cost, and side effects of such efforts pose major problems. We focus here instead on the scientific issues that may help us anticipate our future challenges.

The air, whose composition is a product of biological and industrial processes and of atmospheric chemistry, is a primary factor for Earth's climate and our health and quality of life. This system can only be effectively studied within the broader context of the biogeochemical cycles of carbon, sulfur, and nitrogen, all of which have been perturbed by human activity.

We've already noted how variations of atmospheric CO_2 are linked with such life-defining phenomena as glacial-interglacial cycles and surface temperature. Although the variation in solar insolation caused by orbital changes is the basic forcing mechanism for the glacial-interglacial cycles (as originally suggested by James Croll in 1864 and showed quantitatively by Milutin Milankovitch in 1920 with the aid of a climate model), numerous amplifying mechanisms are required to account for the magnitude of the changes. The CO_2 concentration during the last major glaciation was about 200 ppm and increased to about 280 ppm prior to human CO_2 contributions. Its concentration in 1999 was about 367 ppm, and it continues to rise by about 1.5 ppm annually. Much is now understood about the carbon cycle, but knowledge gaps remain. One of them is the role of terrestrial biota, which have been acting as a sink for CO_2, but may not necessarily play that role indefinitely.

The sulfur cycle also has been perturbed by human activities. Cloud droplets and ice crystals form by nucleation around submicrometer aerosols consisting of sulfates and organics. That is why aerosols and clouds are strongly linked with climate. Human activities have led to a large increase in these aerosol species. The anthropogenic emission of SO_2, which converts into sulfate particles in the atmosphere, exceeds that from natural sources, including volcanic emissions and biological "exhalations," by more than a factor of two.

Sulfate particles cool climate directly by scattering sunlight back into space. Sulfates also cool indirectly by nucleating more cloud drops and increasing the brightness (albedo) of clouds. In 1990, Robert Charlson, Joachim Langer, and Henning Rodhe used a global model of the sulfur budget and estimated that the direct cooling effect of sulfates for the past century may have counteracted as much as 20 to 30 percent of the greenhouse warming. Meanwhile, estimates of cooling from the indirect effect of sulfate range from negligible to offsetting the entire greenhouse forcing.[14]

Carbonaceous aerosols from fossil fuel combustion and biomass burning have become another major source of particles. The combination of strong absorption of light by black carbon and scattering by organics and sulfates significantly reduces the amount of solar energy reaching the surface. That raises the specter of regional effects on the hydrological cycle and photosynthesis. One of the main challenges now is to unravel the linkages between aerosols, cloud formation, and climate change. These include the biogenic sources of aerosols such as dimethyl sulfide from plankton and organics from the land; the anthropogenic addition to the cloud condensation nuclei (CCN); the dependence of cloud cover on aerosols and CCN; the response of water vapor and cloudiness to global warming; and the link between solar absorption by black carbon and the hydrological cycle. Unraveling this Gordian knot must start by connecting processes at the microscopic level of aerosol formation with larger-scale cloud physical and dynamical processes.[16]

We anticipate great advances in our ability to integrate atmospheric phenomena with those of the oceans, land, biosphere, and cryosphere. Global models already have started doing this, but they must be vetted and refined with observations. Atmospheric scientists have a long history of mounting international field experiments, starting with the Global Atmospheric Research Program (conducted by Joachim Kuettner) during the 1970s. Recently, such experiments have begun to integrate physics, chemistry, and dynamics along with data from satellite observations, aircraft, and surface platforms. Only in this way is it possible to begin to grasp the complexity at which atmospheric processes unfold.

One example of this multiscale approach is 1999's Indian Ocean Experiment. The study ranged from the scale of particles to that of an entire ocean basin. It tracked how pollutants from South Asia formed a sulfate-organic-soot haze, which spread over the northern Indian Ocean and significantly reduced the solar heating of the oceans. New satellite platforms, including NASA's TERRA and the European ENVISAT, also have begun to simultaneously measure several vital chemical, biological, and physical parameters of the atmosphere, ocean, and land.

We should also learn a valuable lesson from history. Aristotle's models of winds and of cloud and rain formation survived nearly 2,000 years for lack of observations proving the inadequacy of those models. Due to "prohibitive costs," we too currently lack adequate global-scale observations of such basic atmospheric parameters as the three-dimensional distribution of clouds, aerosols, water vapor, ozone, other chemicals, temperature, and winds. It is our hope that futuristic yet potentially near-term innovations will enable us in the coming decades to float inexpensive microobservers around the globe to gather the data we need to test our ideas, validate our models, and take informed actions.

References and Notes

1. For a general review of the past and future of tropospheric chemistry, see articles in *Science* 276, 1043–1083 (16 May 1997).
2. W. Ramsay, *The Gas of the Atmosphere* (MacMillan and Co., New York, ed. 4, 1915).
3. M. Schmidt, *Pioneers of Ozone Research* (Mecke Druck, Göttingen, Germany, 1988).
4. P. Crutzen, Les Prix Nobel 1995, 116–176. Printed in Sweden (1995).
5. S. Solomon, Stratospheric Ozone Depletion: A Review of Concepts and History, *Rev. of Geophysics,* p. 275 (1999).
6. H. Frisinger, *The History of Meteorology to 1800* (American Meteorological Society, New York, 1977), p. 148.
7. J. R. Fleming, Ed., *Historical Essays on Meteorology 1919–1995* (American Meteorological Society, New York, 1996). See articles by E. Lorenz, G. P. Cressman, and J. Kutzbach.
8. E. N. Lorenz, *The Nature and Theory of the General Circulation of the Atmosphere* (WMO Press, Geneva, 1967), p. 161.
9. M. Wuertele, Ed., *Selected Papers of Jacob Aall Bonnevie Bjerkenes* (Western Periodicals Co., North Hollywood, CA, 1975). See contributions by J. Charney, Y. Mintz, and J. Namias.
10. H. Rodhe, R. Charlson, E. Crawford, Eds., *Svante Arrhenius and the Greenhouse Effect, The Legacy of Svante Arrhenius Understanding the Greenhouse Effect* (Royal Swedish Academy of Sciences, Stockholm, 1997).
11. A. R. Ravishankara, *Science* 276, 1058 (1997).
12. W. B. DeMore et al., Chemical Kinetics and Photochemical Data for Use in Stratospheric Modeling, NASA Jet Propulsion Laboratory, Pasadena, California, 15 January 1997.
13. V. Ramanathan et al., Trace Gas Effects on Climate: Atmospheric Ozone 1985 report by WMO, Geneva, Switzerland. See Chapter 15, Vol. III (1986).
14. J. Houghton et al., Eds., IPCC, Climate Change 1995, The Science of Climate Change (Cambridge University Press, Cambridge, 1995).
15. G. Philander, *El Niño, La Niña and the Southern Oscillation* (Academic Press, San Diego, 1990).
16. B. J. Mason, *The Physics of Clouds* (Clarendon Press, Oxford, U.K., 1971).
17. P. V. Hobbs, *Ice Physics* (Clarendon Press, Oxford, UK, 1974).

18. Thanks go to James R. Fleming of Colby College for assistance with the timeline.

11

Neuroscience: Breaking Down Scientific Barriers to the Study of Brain and Mind

ERIC R. KANDEL AND LARRY R. SQUIRE

Brains and minds are the most fascinating of all phenomena. After all, in a unique self-reflexivity, it is with their own brains and minds that neuroscientists strive to understand brains and minds. And as revealed by Eric R. Kandel (who heard he had won the 2000 Nobel Prize in Medicine or Physiology while working on this chapter) and his long-time colleague and co-author Larry R. Squire, an enormous amount is now known about our brains and minds. These authors describe how brain research has migrated from the peripheries of biology and psychology to assume a central position within those disciplines. The multidiscipline of neuroscience that emerged from this process now encompasses an awesome range of subjects, from genes to cognition, from molecules to minds.

ERIC R. KANDEL is University Professor at Columbia University in New York City and a Senior Investigator at the Howard Hughes Medical Institute. A past president of the Society for Neuroscience, he is a member of the National Academy of Sciences and the Institute of Medicine. In 2000, he was named a co-recipient of the Nobel Prize in Physiology or Medicine. LARRY R. SQUIRE is Research Career Scientist at the Veterans Affairs San Diego Healthcare System and Professor of Psychiatry, Neurosciences, and Psychology at the University of California, San Diego. He too was a past president of the Society of Neuroscience and is now a member of the National Academy of Sciences and the Institute of Medicine.

TIMELINE OF NEUROSCIENCE

Second century A.D.
Galen of Pergamum identifies the brain as the organ of the mind.

Seventeenth century
The brain becomes accepted as the substrate of mental life rather than its ventricles, as early writers had proposed.

1664
Thomas Willis publishes *Cerebri anatome*, with illustrations of the brain by Christopher Wren. It is the most comprehensive treatise on brain anatomy and function published up to that time.

1791
Luigi Galvani reveals the electric nature of nervous action by stimulating nerves and muscles of frog legs.

1808
Franz Joseph Gall proposes that specific brain regions control specific functions.

During the latter part of the twentieth century, the study of the brain moved from a peripheral position within both the biological and psychological sciences to become an interdisciplinary field called neuroscience that now occupies a central position within each discipline. This realignment occurred because the biological study of the brain became incorporated into a common framework with cell and molecular biology on the one side and with psychology on the other. Within this new framework, the scope of neuroscience ranges from genes to cognition, from molecules to mind.

What led to the gradual incorporation of neuroscience into the central core of biology and to its alignment with psychology? From the perspective of biology at the beginning of the twentieth century, the task of neuroscience—to understand how the brain develops and then functions to perceive, think, move, and remember—seemed impossibly difficult. In addition, an intellectual barrier separated neuroscience from biology, because the language of neuroscience was based more on neuroanatomy and electrophysiology than on the universal biological language of biochemistry. During the last two decades this barrier has been largely removed. A molecular neuroscience became established by focusing on simple systems where anatomy and physiology were tractable. As a result, neuroscience helped delineate a general plan for neural cell function in which the cells of the nervous system are understood to be governed by variations on universal biological themes.

From the perspective of psychology, a neural approach to mental processes seemed too reductionistic to do justice to the complexity of cognition. Substantial progress was required to demonstrate that some of these reductionist goals were achievable within a psychologically meaningful framework. The work of Vernon Mountcastle, David Hubel, Torsten Wiesel, and Brenda Milner in the 1950s and 1960s, and the advent of brain imaging in the 1980s, showed what could be achieved for sensory processing, perception, and memory. As a result of these advances, the view gradually developed that only by exploring the brain could psychologists fully satisfy their interest in the cognitive processes that intervene between stimulus and response.

Here, we consider several developments that have been particularly important for the maturation of neuroscience and for the restructuring of its relationship to biology and psychology.

The Emergence of a Cellular and Molecular Neuroscience

The modern cellular science of the nervous system was founded on two important advances: the neuron doctrine and the ionic hypothesis. The neuron doctrine was established by the brilliant Spanish anatomist Santiago Ramón y Cajal,[1] who showed that the brain is composed of discrete cells, called neurons, and that these likely serve as elementary signaling units. Cajal also advanced the principle of connection specificity, the central tenet of which is that neurons form highly specific connections with one another and that these connections are invariant and defining for each species. Finally, Cajal developed the principle of dynamic polarization, according to which information flows in only one direction within a neuron, usually from the dendrites (the neuron's input component) down the axon shaft to the axon terminals (the output component). Although exceptions to this principle have emerged, it has proved extremely influential, because it tied structure to function and provided guidelines for constructing circuits from the images provided in histological sections of the brain.

Seeing neurons. Anatomist Ramón y Cajal used Golgi's stain to examine individual nerve cells and their processes. *Library of Medicine.*

1852
Hermann von Helmholtz measures the speed of a nerve impulse in the frog.

1879
Wilhelm Wundt establishes the first laboratory of experimental psychology in Leipzig, Germany.

1891
Wilhelm von Waldeyer-Hartz introduces the term neuron.

1897
Charles Sherrington introduces the term synapse.

1898–1903
Edward Thorndike and Ivan Pavlov describe operant and classical conditioning, two fundamental types of learning.

1906
Santiago Ramón y Cajal summarizes compelling evidence for the neuron doctrine, that the nervous system is composed of discrete cells.

Alois Alzheimer describes the pathology of the neurodegenerative disease that comes to bear his name.

1914
Henry Dale demonstrates the physiological action of acetylcholine, which is later identified as a neurotransmitter.

1929
In a famous program of lesion experiments in rats, Karl Lashley attempts to localize memory in the brain.
Hans Berger uses human scalp electrodes to demonstrate electroencephalography.

1928–32
Edgar Adrian describes method for recording from single sensory and motor axons; H. Keffer Hartline applies this method to the recording of single-cell activity in the eye of the horseshoe crab.

1940s
Alan Hodgkin, Andrew Huxley, and Bernard Katz explain electrical activity of neurons by concentration gradients of ions and movement of ions through pores.

Cajal and his contemporary Charles Sherrington[2] further proposed that neurons contact one another only at specialized points called synapses, the sites where one neuron's processes contact and communicate with another neuron. We now know that at most synapses, there is a gap of 20 nm—the synaptic cleft—between the pre- and postsynaptic cell. In the 1930s, Otto Loewi, Henry Dale, and Wilhelm Feldberg established (at peripheral neuromuscular and autonomic synapses) that the signal that bridges the synaptic cleft is usually a small chemical, or neurotransmitter, which is released from the presynaptic terminal, diffuses across the gap, and binds to receptors on the postsynaptic target cell. Depending on the specific receptor, the postsynaptic cell can either be excited or inhibited. It took some time to establish that chemical transmission also occurs in the central nervous system, but by the 1950s the idea had become widely accepted.

Even early in the twentieth century, it was already understood that nerve cells have an electrical potential, the resting membrane potential, across their membrane, and that signaling along the axon is conveyed by a propagated electrical signal, the action potential, which was thought to nullify the resting potential. In 1937 Alan Hodgkin discovered that the action potential gives rise to local current flow on its advancing edge and that this current depolarizes the adjacent region of the axonal membrane sufficiently to trigger a traveling wave of depolarization. In 1939 Hodgkin and Andrew Huxley made the surprising discovery that the action potential more than nullifies the resting potential—it reverses it. Then, in the late 1940s, Hodgkin, Huxley, and Bernard Katz explained the resting potential and the action potential in terms of the movement of specific ions—potassium (K^+), sodium (Na^+), and chloride (Cl^-)—through pores (ion channels) in the axonal membrane. This ionic hypothesis unified a large body of descriptive data and offered the first realistic promise that the nervous system could be understood in terms of physicochemical principles common to all of cell biology.[3]

The next breakthrough came when Katz, Paul Fatt, and John Eccles showed that ion channels are also fundamental to signal transmission across the synapse. However, rather than being gated by voltage like the Na^+ and K^+ channels critical for action potentials, excitatory synaptic ion channels are gated chemically by ligands such as the transmitter acetylcholine. During the 1960s and 1970s, neuroscientists identified many amino acids, peptides, and other small molecules as chemical transmitters, including acetylcholine, glutamate, GABA, glycine, serotonin, dopamine, and norepinephrine. On the order of 100 chemical transmitters have been discovered to date. In the 1970s, some synapses were found to release a peptide cotransmitter that can modify the action of the classic, small-molecule transmitters. The discovery of chemical neurotransmission was followed by the remarkable discovery that transmission between neurons is sometimes electrical.[4] Electrical synapses have smaller synaptic clefts, which are bridged by gap junctions and allow current to flow between neurons.

In the late 1960s information began to become available about the biophysical and biochemical structure of ionic pores and the biophysical basis for their selectivity and gating—how they open and close. For example, transmitter binding sites and their ion channels were found to be embodied within different domains of multimeric proteins. Ion channel selectivity was found to depend on physical-chemical interaction between the channel and the ion, and channel gating was found to result from conformational changes within the channel.[5]

The study of ion channels changed radically with the development of the patch-clamp method in 1976 by Erwin Neher and Bert Sakmann,[6] which enabled measurement of the current flowing through a single ion channel. This powerful advance set the stage for the analysis of channels at the molecular level and for the analysis of functional and conformational change in a single membrane protein. When applied to non-neuronal cells, the method also revealed that all cells—even bacteria—express remarkably similar ion channels. Thus, neuronal signaling proved to be a special case of a signaling capability inherent in most cells.

The development of patch clamping coincided with the advent of molecular cloning, and these two methods brought neuroscientists new ideas based on the first reports of the amino acid sequences of ligand- and voltage-gated channels. One of the key insights to emerge from molecular cloning was that amino acid sequences contain clues about how receptor proteins and voltage-gated ion channel proteins are arranged across the cell membrane. The sequence data also often pointed to unexpected structural relationships (homologies) among proteins. These insights, in turn, revealed similarities between molecules found in quite different neuronal and non-neuronal contexts, suggesting that they may serve similar biological functions.

By the early 1980s, it became clear that synaptic actions were not always mediated directly by ion channels. Besides ionotropic receptors, in which ligand binding directly gates an ion channel, a second class of receptors, the metabotropic receptors, was discovered. Here the binding of the ligand initiates intracellular metabolic events and leads only indirectly, by way of "second messengers," to the gating of ion channels.[7]

The cloning of metabotropic receptors revealed that many of them have

1946

Kenneth Cole develops the voltage-clamp technique to measure current flow across the cell membrane.

1949

Donald Hebb introduces a synaptic learning rule, which becomes known as the Hebb rule.

1930s to 1950s

The chemical nature of synaptic transmission is established by Otto Loewi, Henry Dale, Wilhelm Feldberg, Stephen Kuffler, and Bernard Katz at peripheral synapses and is extended to the spinal cord by John Eccles and others.

Wilder Penfield and Theodore Rasmussen map the motor and sensory homunculus and illustrate localization of function in the human brain.

1950s

Karl von Frisch, Konrad Lorenz, and Nikolaas Tinbergen establish the science of ethology (animal behavior in natural contexts) and lay the foundation for neuroethology.

1955–60
Vernon Mountcastle, David Hubel, and Torsten Wiesel pioneer single-cell recording from mammalian sensory cortex; Nils-Ake Hillarp introduces fluorescent microscopic methods to study cellular distribution of biogenic amines.

1956
Rita Levi-Montalcini and Stanley Cohen isolate and purify nerve growth factor.

1957
Brenda Milner describes patient H.M. and discovers the importance of the medial temporal lobe for memory.

1958
Arvid Carlsson finds dopamine to be a transmitter in the brain and proposes that it has a role in extrapyramidal disorders such as Parkinson's disease.

1960s
Simple invertebrate systems, including *Aplysia, Drosophila,* and *C. elegans,* are introduced to analyze elementary aspects of behavior and learning at the cellular and molecular level.

seven membrane-spanning regions and are homologous to bacterial rhodopsin as well as to the photoreceptor pigment of organisms ranging from fruit flies to humans. Further, the recent cloning of receptors for the sense of smell[8] revealed that at least 1000 metabotropic receptors are expressed in the mammalian olfactory epithelium and that similar receptors are present in flies and worms. Thus, it was instantly understood that the class of receptors used for phototransduction, the initial step in visual perception, is also used for smell and aspects of taste, and that these receptors share key features with many other brain receptors that work through second-messenger signaling. These discoveries demonstrated the evolutionary conservation of receptors and emphasized the wisdom of studying a wide variety of experimental systems—vertebrates, invertebrates, even single-celled organisms—to identify broad biological principles.

The seven transmembrane-spanning receptors activate ion channels indirectly through coupling proteins (G proteins). Some G proteins have been found to activate ion channels directly. However, the majority of G proteins activate membrane enzymes that alter the level of second messengers, such as cAMP, cGMP, or inositol triphosphate, which initiate complex intracellular events leading to the activation of protein kinases and phosphatases and then to the modulation of channel permeability, receptor sensitivity, and transmitter release. Neuroscientists now appreciate that many of these synaptic actions are mediated intracellularly by protein phosphorylation or dephosphorylation.[9] Nerve cells use such covalent modifications to control protein activity reversibly and thereby to regulate function. Phosphorylation is also critical in other cells for the action of hormones and growth factors, and for many other processes.

Directly controlled synaptic actions are fast, lasting milliseconds, but second-messenger actions last seconds to minutes. An even slower synaptic action, lasting days or more, has been found to be important for long-term memory. In this case, protein kinases activated by second messengers translocate to the nucleus, where they phosphorylate transcription factors that alter gene expression, initiate growth of neuronal processes, and increase synaptic strength.

Ionotropic and metabotropic receptors have helped to explain the postsynaptic side of synaptic transmission. In the 1950s and 1960s, Katz and his colleagues turned to the presynaptic terminals and discovered that chemical transmitters, such as acetylcholine, are released not as single molecules but as packets of about 5,000 molecules called quanta.[10] Each quantum is packaged in a synaptic vesicle and released by exocytosis at sites called active zones. The key signal that triggers this sequence is the influx of Ca^{2+} with the action potential.

In recent years, many proteins involved in transmitter release have been identified.[11] Their functions range from targeting vesicles to active zones, tethering vesicles to the cell membrane, and fusing vesicles with the cell membrane so that their contents can be released by exocytosis. These molecular studies reflect another example of evolutionary conservation: The molecules used for vesicle fusion and exocytosis at nerve terminals are variants of those used for vesicle fusion and exocytosis in all cells.

A Mechanistic View of Brain Development

The discoveries of molecular neuroscience have dramatically improved the understanding of how the brain develops its complexity. The modern molecular era of developmental neuroscience began when Rita Levi-Montalcini and Stanley Cohen isolated nerve growth factor (NGF), the first peptide growth factor to be identified in the nervous system.[12] They showed that injection of antibodies to NGF into newborn mice caused the death of neurons in sympathetic ganglia and also reduced the number of sensory ganglion cells. Thus, the survival of both sympathetic and sensory neurons depends on NGF. Indeed, many neurons depend for their survival on NGF or related molecules, which typically provide feedback signals to the neurons from their targets. Such signals are important for programmed cell death—apoptosis—a developmental strategy which has now proved to be of general importance, whereby many more cells are generated than eventually survive to become functional units with precise connectivity. In a major advance, genetic study of worms has revealed the *ced* genes and with them a universal cascade critical for apoptosis in which proteases—the caspases—are the final agents for cell death.[13]

Cajal pointed out the extraordinary precision of neuronal connections. The first compelling insights into how neurons develop their precise connectivity came from Roger Sperry's studies of the visual system of frogs and salamanders beginning in the 1940s, which suggested that axon outgrowth is guided by molecular cues. Sperry's key finding was that when the nerves from the eye are cut, axons find their way back to their original targets. These seminal studies led Sperry in 1963 to formulate the chemoaffinity hypothesis,[14] the idea that neurons form connections with their targets based on distinctive and matching molecular identities that they acquire early in development.

1962–63
Brain anatomy in rodents is found to be altered by experience; first evidence for role of protein synthesis in memory formation.

1963
Roger Sperry proposes a precise system of chemical matching between pre- and postsynaptic neuronal partners (the chemoaffinity hypothesis).

1966–69
Ed Evarts and Robert Wurtz develop methods for studying movement and perception with single-cell recordings from awake, behaving monkeys.

1970
Synaptic changes are related to learning and memory storage in *Aplysia*.

Mid-1970s
Paul Greengard shows that many neurotransmitters work by means of protein phosphorylation.

1973
Timothy Bliss and Terje Lomo discover long-term potentiation, a candidate synaptic mechanism for long-term mammalian memory.

1976
Erwin Neher and Bert Sakmann develop the patch-clamp technique for recording the activity of single ion channels.

Late 1970s
Neuroimaging by positron emission tomography is developed.

1980s
Experimental evidence becomes available for the divisibility of memory into multiple systems; an animal model of human amnesia is developed.

1986
H. Robert Horvitz discovers the *ced* genes, which are critical for programmed cell death.
Patient R.B. establishes the importance of the hippocampus for human memory.

Stimulated by these early contributions, molecular biology has radically transformed the study of nervous system development from a descriptive to a mechanistic field. Three genetic systems, the worm *Caenorhabditis elegans,* the fruit fly *Drosophila melanogaster,* and the mouse, have been invaluable; some of the molecules for key developmental steps in the mouse were first characterized by genetic screens in worms and flies. In some cases, identical molecules were found to play an equivalent role throughout phylogeny. The result of this work is that neuroscientists have achieved in broad outline an understanding of the molecular basis of nervous system development.[15] A range of key molecules has been identified, including specific inducers, morphogens, and guidance molecules important for differentiation, process outgrowth, pathfinding, and synapse formation. For example, in the spinal cord, neurons achieve their identities and characteristic positions largely through two classes of inductive signaling molecules of the Hedgehog and bone morphogenic protein families. These two groups of molecules control neuronal differentiation in the ventral and dorsal halves of the spinal cord, respectively, and maintain this division of labor through most of the rostrocaudal length of the nervous system.

The process of neuronal pathfinding is mediated by both short-range and long-range cues. An axon's growth cone can encounter cell surface cues that either attract or repel it. For example, ephrins are membrane-bound, are distributed in graded fashion in many regions of the nervous system, and can repel growing axons. Other cues, such as the netrins and the semaphorins, are secreted in diffusible form and act as long-range chemoattractants or chemorepellents. Growth cones can also react to the same cues differently at different developmental phases, for example, when crossing the midline or when switching from pathfinding to synapse formation. Finally, a large number of molecules are involved in synapse formation itself. Some, such as neuregulin, erbB kinases, agrin, and MuSK, organize the assembly of the postsynaptic machinery, whereas others, such as the laminins, help to organize the presynaptic differentiation of the active zone.

These molecular signals direct differentiation, migration, process outgrowth, and synapse formation in the absence of neural activity. Neural activity is needed, however, to refine the connections further so as to forge the adult pattern of connectivity.[16] The neural activity may be generated spontaneously, especially early in development, but later depends importantly on sensory input. In this way, intrinsic activity or sensory and motor experience can help specify a precise set of functional connections.

The Impact of Neuroscience on Neurology and Psychiatry

Molecular neuroscience has also reaped substantial benefits for clinical medicine. To begin with, recent advances in the study of neural development have identified stem cells, both embryonic and adult, which offer promise in cell replacement therapy in Parkinson's disease, demyelinating diseases, and other conditions. Similarly, new insights into axon guidance molecules offer hope for nerve regeneration after spinal cord injury. Finally, because most neurological diseases are associated with cell death, the discovery in worms of a universal genetic program for cell death opens up approaches for cell rescue based on, for example, inhibition of the caspase proteases.

Next, consider the impact of molecular genetics. Huntington's disease is an autosomal dominant disease marked by progressive motor and cognitive impairment that ordinarily manifests itself in middle age. The major pathology is cell death in the basal ganglia. In 1993, the Huntington's Disease Collaborative Research Group isolated the gene responsible for the disease.[17] It is marked by an extended series of trinucleotide CAG (cytosine, adenine, guanine) repeats, thereby placing Huntington's disease in a new class of neurological disorders—the trinucleotide repeat diseases—that now constitute the largest group of dominantly transmitted neurological diseases.

The molecular genetic analysis of more complex degenerative disorders has proceeded more slowly. Still, three genes associated with familial Alzheimer's disease—those that code for the amyloid precursor protein, presenilin 1, and presenilin 2—have been identified. Molecular genetic studies have also identified the first genes that modulate the severity and risk of a degenerative disease.[18] One allele (APO E4) is a significant risk factor for late-onset Alzheimer's disease. Conversely, the APO E2 allele may actually be protective. A second risk factor is a_2-macroglobulin. All the Alzheimer's-related genes so far identified participate in either generating or scavenging a protein (the amyloid peptide), which is toxic at elevated levels. Studies directed at this peptide may lead to ways to prevent the disease or halt its progression. Similarly, the discovery of b-secretase and perhaps g-secretase, the enzymes involved in the processing of b amyloid, represent dramatic advances that may also lead to new treatments.

With psychiatric disorders, progress has been slower for two reasons. First, diseases such as schizophrenia, depression, obsessive-compulsive disorders, anxiety states, and drug abuse tend to be complex, polygenic disorders that are significantly modulated by environmental factors. Second, in contrast to neurological disorders, little is known about the anatomical substrates of most

1990s
Neural development is transformed from a descriptive to a molecular discipline by Gerald Fischbach, Jack McMahan, Tom Jessell, and Corey Goodman; neuroimaging is applied to problems of human cognition, including perception, attention, and memory.
Reinhard Jahn, James Rothman, Richard Scheller, and Thomas Sudhof delineate molecules critical for exocytosis.

1990
Segi Ogawa and colleagues develop functional magnetic resonance imaging.
Mario Capecchi and Oliver Smythies develop gene knockout technology, which is soon applied to neuroscience.

1991
Linda Buck and Richard Axel discover that the olfactory receptor family consists of over 1,000 different genes. The anatomical components of the medial temporal lobe memory system are identified.

1993
The Huntington's Disease Collaborative Research Group identifies the gene responsible for Huntington's disease.

1998
First 3D structure of an ion channel is revealed by Rod MacKinnon.

psychiatric diseases. Given the difficulty of penetrating the deep biology of mental illness, it is nevertheless remarkable how much progress has been made during the past three decades.[19] Arvid Carlsson and Julius Axelrod carried out pioneering studies of biogenic amines, which laid the foundation for psychopharmacology, and Seymour Kety pioneered the genetic study of mental illness.[20] Currently, new approaches to many conditions, such as sleep disorders, eating disorders, and drug abuse, are emerging as the result of insights into the cellular and molecular machinery that regulates specific behaviors.[21] Moreover, improvements in diagnosis, the better delineation of genetic contributions to psychiatric illness (based on twin and adoption studies as well as studies of affected families), and the discovery of specific medications for treating schizophrenia, depression, and anxiety states have transformed psychiatry into a therapeutically effective medical specialty that is now closely aligned with neuroscience.

A New Alignment of Neuroscience and Psychological Science

The brain's computational power is conferred by interactions among billions of nerve cells, which are assembled into networks or circuits that carry out specific operations in support of behavior and cognition. Whereas the molecular machinery and electrical signaling properties of neurons are widely conserved across animal species, what distinguishes one species from another, with respect to their cognitive abilities, is the number of neurons and the details of their connectivity.

Beginning in the nineteenth century there was great interest in how these cognitive abilities might be localized in the brain. One view, first championed by Franz Joseph Gall, was that the brain is composed of specialized parts and that aspects of perception, emotion, and language can be localized to anatomically distinct neural systems. Another view, championed by Jean-Pierre-Marie Flourens, was that cognitive functions are global properties arising from the integrated activity of the entire brain. In a sense, the history of neuroscience can be seen as a gradual ascendancy of the localizationist view.

To a large extent, the emergence of the localizationist view was built on a century-old legacy of psychological science. When psychology emerged as an experimental science in the late nineteenth century, its founders, Gustav Fechner and Wilhelm Wundt, focused on psychophysics—the quantitative relationship between physical stimuli and subjective sensation. The success of this endeavor encouraged psychologists to study more complex behavior, which led to a rigorous, laboratory-based tradition termed behaviorism.

Led by John Watson and later by B. F. Skinner, behaviorists argued that psychology should be concerned only with observable stimuli and responses, not with unobservable processes that intervene between stimulus and response. This tradition yielded lawful principles of behavior and learning, but it proved limiting. In the 1960s, behaviorism gave way to a broader approach concerned with cognitive processes and internal representations. This new emphasis focused on precisely those aspects of mental life—from perception to action—that had long been of interest to neurologists and other students of the nervous system.

The first cellular studies of brain systems in the 1950s illustrated dramatically how much neuroscience derived from psychology and conversely how much psychology could, in turn, inform neuroscience. In using a cellular approach, neuroscientists relied on the rigorous experimental methods of psychophysics and behaviorism to explore how a sensory stimulus resulted in a neuronal response. In so doing, they found cellular support for localization of function: Different brain regions had different cellular response properties. Thus, it became possible in the study of behavior and cognition to move beyond description to an exploration of the mechanisms underlying the internal representation of the external world.

In the late 1950s and 1960s Mountcastle, Hubel, and Wiesel began using cellular approaches to analyze sensory processing in the cerebral cortex of cats and monkeys.[22] Their work provided the most fundamental advance in understanding the organization of the brain since the work of Cajal at the turn of the century. The cellular physiological techniques revealed that the brain both filters and transforms sensory information on its way to and within the cortex, and that these transformations are critical for perception. Sensory systems analyze, decompose, and then restructure raw sensory information according to built-in connections and rules.

Mountcastle found that single nerve cells in the primary somatic sensory cortex respond to specific kinds of touch: Some respond to superficial touch and others to deep pressure, but cells almost never respond to both. The different cell types are segregated in vertical columns, which comprise thousands of neurons and extend about 2 mm from the cortical surface to the white matter below it. Mountcastle proposed that each column serves as an integrating unit, or logical module, and that these columns are the basic mode of cortical organization.

Single-cell recording was pioneered by Edgar Adrian and applied to the visual system of invertebrates by H. Keffer Hartline and to the visual system of mammals by Stephen Kuffler, the mentor of Hubel and Wiesel. In recordings from the retina, Kuffler discovered that, rather than signaling absolute levels of light, neurons signal contrast between spots of light and dark. In the visual cortex, Hubel and Wiesel found that most cells no longer respond to spots of light. For example, in area V1 at the occipital pole of the cortex, neurons respond to specific visual features such as lines or bars in a particular orientation. Moreover, cells with similar orientation preferences were found to group together in vertical columns similar to those that Mountcastle had found in somatosensory cortex. Indeed, an independent system of vertical columns—the ocular dominance columns—was found to segregate information arriving from the two eyes. These results provided an entirely new view of the anatomical organization of the cerebral cortex.

Wiesel and Hubel also investigated the effects of early sensory deprivation on newborn animals. They found that visual deprivation in one eye profoundly alters the organization of ocular dominance columns.[23] Columns receiving input from the closed eye shrink, and those receiving input from the open eye expand. These studies led to the discovery that eye closure alters the pattern of synchronous activity in the two eyes and that this neural activity is essential for fine-tuning synaptic connections during visual system development.[16]

In the extrastriate cortex beyond area V1, continuing electrophysiological and anatomical studies have identified more than 30 distinct areas important for vision.[24] Further, visual information was found to be analyzed by two parallel processing streams.[25] The dorsal stream, concerned with where objects are located in space and how to reach objects, extends from area V1 to the parietal cortex. The ventral stream extends from area V1 to the inferior temporal cortex and is concerned with analyzing the visual form and quality of objects. Thus, even the apparently simple task of perceiving an object in space engages a disparate collection of specialized neural areas that represent different aspects of the visual information—what the object is, where it is located, and how to reach for it.

A Neuroscience of Cognition

The initial studies of the visual system were performed in anaesthetized cats, an experimental preparation far removed from the behaving and thinking human beings that are the focus of interest for cognitive psychologists. A pivotal advance occurred in the late 1960s when single-neuron recordings were obtained from awake, behaving monkeys that had been trained to perform sensory or motor tasks.[26] With these methods, the response of neurons in the posterior parietal cortex to a visual stimulus was found to be enhanced when the animal moved its eyes to attend to the stimulus. This moved the neurophysiological study of single neurons beyond sensory processing and showed that reductionist approaches could be applied to higher-order psychological processes such as selective attention.

It is possible to correlate neuronal firing with perception rather directly. Thus, building on earlier work by Mountcastle, a monkey's ability to discriminate motion was found to closely match the performance of individual neurons in area MT, a cortical area concerned with visual motion processing. Further, electrical microstimulation of small clusters of neurons in MT shifts the monkey's motion judgments toward the direction of motion that the stimulated neurons prefer.[27] Thus, activity in area MT appears sufficient for the perception of motion and for initiating perceptual decisions.

These findings, based on recordings from small neuronal populations, have illuminated important issues in perception and action. They illustrate how retinal signals are remapped from retinotopic space into other coordinate frames that can guide behavior; how attention can modulate neuronal activity; and how meaning and context influence neuronal activity, so that the same retinal stimulus can lead to different neuronal responses depending on how the stimulus is perceived.[28] This same kind of work (relating cellular activity directly to perception and action) is currently being applied to the so-called binding problem—how the multiple features of a stimulus object, which are represented by specialized and distributed neuronal groups, are synthesized into a signal that represents a single percept or action and to the fundamental question of what aspects of neuronal activity (e.g., firing rate or spike timing) constitute the neural codes of information processing.[29]

Striking parallels to the organization and function of sensory cortices have been found in the cortical motor areas supporting voluntary movement. Thus, there are several cortical areas directed to the planning and execution of voluntary movement. Primary motor cortex has columnar organization, with neurons in each column governing movements of one or a few joints. Motor areas receive input from other cortical regions, and information moves through stages to the spinal cord, where the detailed circuitry that generates motor patterns is located.[30]

Although studies of single cells have been enormously informative, the functioning brain consists of multiple brain systems and many neurons operating in concert. To monitor activity in large populations of neurons, multielectrode arrays as well as cellular and whole-brain imaging techniques are now being used. These approaches are being supplemented by studying the effect of selective brain lesions on behavior and by molecular methods, such as the delivery of markers or other molecules to specific neurons by viral transfection, which promise fine-resolution tracing of anatomical connections, activity-dependent labeling of neurons, and ways to transiently inactivate specific components of neural circuits.

Invasive molecular manipulations of this kind cannot be applied to humans. However, functional neuroimaging by positron emission tomography (PET) or functional magnetic resonance imaging (fMRI) provides a way to monitor large neuronal populations in awake humans while they engage in cognitive tasks.[31] PET involves measuring regional blood flow using $H_2\,^{15}O$ and allows for repeated measurements on the same individual. fMRI is based on the fact that neural activity changes local oxygen levels in tissue and that oxygenated and deoxygenated hemoglobin have different magnetic properties. It is now possible to image the second-by-second time course of the brain's response to single stimuli or single events with a spatial resolution in the millimeter range. Recent success in obtaining fMRI images from awake monkeys, combined with single-cell recording, should extend the utility of functional neuroimaging by permitting parallel studies in humans and nonhuman primates.

One example of how parallel studies of humans and nonhuman primates have advanced the understanding of brain systems and cognition is in the study of memory. The neuroscience of memory came into focus in the 1950s when the noted amnesic patient H.M. was first described.[32] H.M. developed profound forgetfulness after sustaining a bilateral medial temporal lobe resection to relieve severe epilepsy. Yet he retained his intelligence, perceptual abilities, and personality. Brenda Milner's elegant studies of H.M. led to several important principles. First, acquiring new memories is a distinct cerebral function, separable from other perceptual and cognitive abilities. Second, because H.M. could retain a number or a visual image for a short time, the medial temporal lobes are not needed for immediate memory. Third, these structures are not the ultimate repository of memory, because H.M. retained his remote, childhood memories.

It subsequently became clear that only one kind of memory, declarative memory, is impaired in H.M. and other amnesic patients. Thus, memory is not a unitary faculty of the mind but is composed of multiple systems that have different logic and neuroanatomy.[33] The major distinction is between our capacity for conscious, declarative memory about facts and events and a collection of unconscious, nondeclarative memory abilities, such as skill and habit learning and simple forms of conditioning and sensitization. In these cases, experience modifies performance without requiring any conscious memory content or even the experience that memory is being used.

An animal model of human amnesia in the nonhuman primate was achieved in the early 1980s, leading ultimately to the identification of the medial temporal lobe structures that support declarative memory—the hippocampus and the adjacent entorhinal, perirhinal, and parahippocampal cortices.[34] The hippocampus has been an especially active target of study, in part because this was one of the structures damaged in patient H.M. and also because of the early discovery of hippocampal place cells, which signal the location of an animal in space.[35] This work led to the idea that, once learning occurs, the hippocampus and other medial temporal lobe structures permit the transition to long-term memory, perhaps by binding the separate cortical regions that together store memory for a whole event. Thus, long-term memory is thought to be stored in the same distributed set of cortical structures that perceive, process, and analyze what is to be remembered, and aggregate changes in large assemblies of cortical neurons are the substrate of long-term memory. The frontal lobes are also thought to influence what is selected for storage, the ability to hold information in mind for the short term, and the ability later on to retrieve it.[36]

Whereas declarative memory is tied to a particular brain system, nondeclarative memory refers to a collection of learned abilities with different brain substrates. For example, many kinds of motor learning depend on the cerebellum, emotional learning and the modulation of memory strength by emotion depend on the amygdala, and habit learning depends on the basal ganglia.[37] These forms of nondeclarative memory, which provide for myriad unconscious ways of responding to the world, are evolutionarily ancient and observable in simple invertebrates such as *Aplysia* and *Drosophila.* By virtue of the unconscious status of these forms of memory, they create some of the mystery of human experience. For here arise the dispositions, habits, attitudes, and preferences that are inaccessible to conscious recollection, yet are shaped by past events, influence our behavior and our mental life, and are a fundamental part of who we are.

Bridging Cognitive Neuroscience and Molecular Biology in the Study of Memory Storage

The removal of scientific barriers at the two poles of the biological sciences—in the cell and molecular biology of nerve cells on the one hand, and in the biology of cognitive processes on the other—has raised the question: Can one anticipate an even broader unification, one that ranges from molecules to mind? A beginning of just such a synthesis may be apparent in the study of synaptic plasticity and memory storage.

For all of its diversity, one can view neuroscience as being concerned with two great themes—the brain's "hard wiring" and its capacity for plasticity. The former refers to how connections develop between cells, how cells function and communicate, and how an organism's inborn functions are organized—its sleep-wake cycles, hunger and thirst, and its ability to perceive the world. Thus, through evolution the nervous system has inherited many adaptations that are too important to be left to the vagaries of individual experience. In contrast, the capacity for plasticity refers to the fact that nervous systems can adapt or change as the result of the experiences that occur during an individual lifetime. Experience can modify the nervous system, and as a result organisms can learn and remember.

The precision of neural connections poses deep problems for the plasticity of behavior. How does one reconcile the precision and specificity of the brain's wiring with the known capability of humans and animals to acquire new knowledge? And how is knowledge, once acquired, retained as long-term memory? A key insight about synaptic transmission is that the precise connections between neurons are not fixed but are modifiable by experience. Beginning in 1970, studies in invertebrates such as *Aplysia* showed that simple forms of learning—habituation, sensitization, and classical conditioning—result in functional and structural changes at synapses between the neurons that mediate the behavior being modified. These changes can persist for days or weeks and parallel the time course of the memory process.[38] These cell biological studies have been complemented by genetic studies in *Drosophila*. As a result, studies in *Aplysia* and *Drosophila* have identified a number of proteins important for memory.[39]

In his now-famous book, *The Organization of Behavior,* Donald Hebb proposed in 1949 that the synaptic strength between two neurons should increase when the neurons exhibit coincident activity.[40] In 1973, a long-lasting synaptic plasticity of this kind was discovered in the hippocampus (a key structure for declarative memory).[41] In response to a burst of high-frequency stimuli, the major synaptic pathways in the hippocampus undergo a long-term change, known as long-term potentiation or LTP. The advent in the 1990s of the ability to genetically modify mice made it possible to relate specific genes both to synaptic plasticity and to intact animal behavior, including memory. These techniques now allow one to delete specific genes in specific brain regions and also to turn genes on and off. Such genetic and pharmacological experiments in intact animals suggest that interference with LTP at a specific synapse—the Schaffer collateral-CA1 synapse—commonly impairs memory for space and objects. Conversely, enhancing LTP at the same synapse can enhance memory in these same declarative memory tasks. The findings emerging from these new methods[42] complement those in *Aplysia* and *Drosophila* and reinforce one of Cajal's most prescient ideas: Even though the anatomical connections between neurons develop according to a definite plan, their strength and effectiveness are not predetermined and can be altered by experience.

Combined behavioral and molecular genetic studies in *Drosophila, Aplysia,* and mouse suggest that, despite their different logic and neuroanatomy, declarative and nondeclarative forms of memory share some common cellular and molecular features. In both systems, memory storage depends on a short-term process lasting minutes and a long-term process lasting days or longer. Short-term memory involves covalent modifications of preexisting proteins, leading to the strengthening of preexisting synaptic connections. Long-term memory involves altered gene expression, protein synthesis, and the growth of new synaptic connections. In addition, a number of key signaling molecules involved in converting transient short-term plasticity to persistent long-term memory appear to be shared by both declarative and nondeclarative memory. A striking feature of neural plasticity is that long-term memory involves structural and functional change.[38, 43] This has been shown most directly in invertebrates and is likely to apply to vertebrates as well, including primates.

It had been widely believed that the sensory and motor cortices mature early in life and thereafter have a fixed organization and connectivity. However, it is now clear that these cortices can be reshaped by experience.[44] In one experiment, monkeys learned to discriminate between two vibrating stimuli applied to one finger. After several thousand trials, the cortical representation of the trained finger became more than twice as large as the corresponding areas for other fingers. Similarly, in a neuroimaging study of right-handed string musicians the cortical representations of the fingers of the left hand (whose fingers are manipulated individually and are engaged in skillful playing) were larger than in nonmusicians. Thus, improved finger skills even involve changes in how sensory cortex represents the fingers. Because all organisms experience a different sensory environment, each brain is modified differently. This gradual creation of unique brain architecture provides a biological basis for individuality.

Coda

Physicists and chemists have often distinguished their disciplines from the field of biology, emphasizing that biology was overly descriptive, atheoretical, and lacked the coherence of the physical sciences. This is no longer quite true. In the twentieth century, biology matured and became a coherent discipline as a result of the substantial achievements of molecular biology. In the second half of the century, neuroscience emerged as a discipline that concerns itself with both biology and psychology and that is beginning to achieve a similar coherence. As a result, fascinating insights into the biology of cells, and remarkable principles of evolutionary conservation, are emerging from the study of nerve cells. Similarly, entirely new insights into the nature of mental processes (perception, memory, and cognition) are emerging from the study of neurons, circuits, and brain systems, and computational studies are providing models that can guide experimental work. Despite this remarkable progress, the neuroscience of higher cognitive processes is only beginning. For neuroscience to address the most challenging problems confronting the behavioral and biological sciences, we will need to continue to search for new molecular and cellular approaches and use them in conjunction with systems neuroscience and psychological science. In this way, we will best be able to relate molecular events and specific changes within neuronal circuits to mental processes such as perception, memory, thought, and possibly consciousness itself.

References and Notes

1. S. Ramón y Cajal, *Nobel Lectures: Physiology or Medicine* (1901-1921) (Elsevier, Amsterdam, 1967), pp. 220–253.
2. C. S. Sherrington, *The Central Nervous System,* vol. 3 of *A Textbook of Physiology,* M. Foster, Ed. (MacMillan, London, ed. 7, 1897).
3. A. L. Hodgkin and A. F. Huxley, *Nature* 144, 710 (1939); A. L. Hodgkin et al., *J. Physiol. (Lond.)* 116, 424 (1952); A. L. Hodgkin and A. F. Huxley, *J. Physiol. (Lond.)* 117, 500 (1952).
4. E. J. Furshpan and D. D. Potter, *Nature* 180, 342 (1957); M. V. L. Bennett, in *Structure and Function of Synapses,* G. D. Pappas and D. P. Purpura, Eds. (Raven Press, New York, 1972), pp. 221–256.
5. C. M. Armstrong and B. Hille, *Neuron* 20, 371 (1998); W. A. Catterall, *Neuron,* in press; B. Hille et al., *Nature Medicine* 5, 1105 (1999); D. A. Doyle et al., *Science* 280, 69 (1998); J. P. Changeux and S. J. Edelstein, *Neuron* 21, 959 (1998); A. Karlin, *Harvey Lecture Series* 85, 71 (1991).
6. E. Neher and B. Sakmann, *Nature* 260, 799 (1976).
7. R. J. Lefkowitz, *Nat. Cell Biol.* 2, E133-6 (2000).
8. L. Buck and R. Axel, *Cell* 65, 175 (1991).
9. E. J. Nestler and P. Greengard, *Protein Phosphorylation in the Nervous System* (Wiley, New York, 1984).
10. J. Del Castillo and B. Katz, *J. Physiol.* 124, 560 (1954); B. Katz, in *The Xth Sherrington Lecture* (Thomas, Springfield, IL, 1969).
11. T. Sudhof, *Nature* 375, 645 (1995); R. Scheller, *Neuron* 14, 893 (1995); J. A. McNew et al., *Nature* 407, 153 (2000).
12. S. Cohen and R. Levi-Montalcini, *Proc. Natl. Acad. Sci. U.S.A.* 42, 571 (1956); W. M. Cowan, *Neuron* 20, 413 (1998).
13. M. M. Metzstein et al., *Trends Genet.* 14, 410 (1998).
14. R. W. Sperry, *Proc. Natl. Acad. Sci. U.S.A.* 50, 703 (1963); R. W. Hunt and W. M. Cowan, in *Brain, Circuits and Functions of Mind,* C. B. Trevarthen, Ed. (Cambridge Univ. Press, Cambridge, 1990), pp. 19–74.
15. M. Tessier-Lavigne and C. S. Goodman, *Science* 274, 1123 (1996); T. M. Jessell, *Nature Rev. Genet.* 1, 20 (2000); T. M. Jessell and J. R. Sanes, *Curr. Opin. Neurobiol.,* in press.
16. L. C. Katz and C. J. Shatz, *Science* 274, 1133 (1996).
17. Huntington's Disease Collaborative Research Group, *Cell* 72, 971 (1993); H. L. Paulson and K. H. Fischbeck, *Annu. Rev. Neurosci.* 19, 79 (1996).
18. D. M. Walsh et al., *Biochemistry* 39, 10831 (2000); W. J. Strittmatter and A. D. Roses, *Annu. Rev. Neurosci.* 19, 53 (1996); D. L. Price, *Nature* 399 (6738) (Suppl), A3-5 (1999).
19. S. E. Hyman, *Arch. Gen. Psychiatr.* 157, 88 (2000); D. Charney et al., Eds., *Neurobiology of Mental Illness* (Oxford, New York, 1999); S. H. Barnodes, *Molecules and Mental Illness* (Scientific American Library, New York, 1993); S. Snyder, *Drugs and the Brain* (Scientific American Library, New York, 1986).
20. A. Carlsson, *Annu. Rev. Neurosci.* 10, 19 (1987); J. Axelrod, *Science* 173, 598 (1971); S. S. Kety, *Am. J. Psychiatr.* 140, 720 (1983); N. A. Hillarp et al., *Pharmacol. Rev.* 18, 727 (1966).
21. T. S. Kilduff and C. Peyron, *Trends Neurosci.* 23, 359 (2000); K. L. Houseknecht et al., *J. Anim. Sci.* 76, 1405 (1998); G. F. Koob, *Ann. N.Y. Acad. Sci.* 909, 17 (2000); E. J. Nestler, *Curr. Opin. Neurobiol.* 7, 713 (1997).
22. V. B. Mountcastle, *J. Neurophysiol.* 20, 408 (1957); D. H. Hubel and T. N. Wiesel, *J. Phys-*

iol. 148, 574 (1959); D. H. Hubel and T. N. Wiesel, *Neuron* 20, 401 (1998).

23. T. Wiesel and D. Hubel, *J. Neurophysiol.* 26, 1003 (1963).

24. D. Van Essen, in *Cerebral Cortex,* A. Peters and E. G. Jones, Eds. (Plenum Publishing Corp., New York, 1985), vol. 3, pp. 259–327; S. Zeki, *Nature* 274, 423 (1978); J. Kaas and P. Garraghty, *Curr. Opin. Neurobiol.* 4, 522 (1992).

25. L. Ungerleider and M. Mishkin, in *The Analysis of Visual Behavior,* D. J. Ingle et al., Eds. (MIT Press, Cambridge, MA, 1982), pp. 549–586; A. Milner and M. Goodale, *The Visual Brain in Action* (Oxford, New York, 1995).

26. R. H. Wurtz, *J. Neurophysiol.* 32, 727 (1969); E. V. Evarts, in *Methods in Medical Research,* R. F. Rushman, Ed. (Year Book, Chicago, 1966), pp. 241–250.

27. W. T. Newsome et al., *Nature* 341, 52 (1989); C. D. Salzman et al., *Nature* 346, 174 (1990).

28. R. A. Andersen et al., *Annu. Rev. Neurosci.* 20, 303 (1997); R. Desimone and J. Duncan, *Annu. Rev. Neurosci.* 18, 193 (1995); M. I. Posner and C. D. Gilbert, *Proc. Natl. Acad. Sci. U.S.A.* 96, 2585 (1999); T. D. Albright and G. R. Stoner, *Proc. Natl. Acad. Sci. U.S.A.* 92, 2433 (1995); C. D. Gilbert, *Physiol. Rev.* 78, 467 (1998); N. K. Logothetis, *Philos. Trans. R. Soc. London, Ser. B* 353, 1801 (1998).

29. For the binding problem, see *Neuron* 24 (1) (1999); for neural codes, see M. N. Shadlen and W. T. Newsome, *Curr. Opin. Neurobiol.* 4, 569 (1994); W. R. Softky, *Curr. Opin. Neurobiol.* 5, 239 (1995).

30. S. Grillner et al., Eds., *Neurobiology of Vertebrate Locomotion,* Wenner-Gren Center International Symposium Series, vol. 45 (Macmillan, London, 1986); A. P. Georgopoulos, *Curr. Opin. Neurobiol.* 10, 238 (2000).

31. L. Sokoloff et al., *J. Neurochem.* 28, 897 (1977); M. Reivich et al., *Circ. Res.* 44 127 (1979); M. I. Posner and M. E. Raichle, *Images of Mind* (Scientific American Library, New York, 1994); S. Ogawa et al., *Proc. Natl. Acad. Sci. U.S.A.* 87, 9868 (1990); B. R. Rosen et al., *Proc. Natl. Acad. Sci. U.S.A.* 95, 773 (1998).

32. W. B. Scoville and B. Milner, *J. Neurol., Neurosurg., Psychiatr.* 20, 11 (1957); B. Milner et al., *Neuron* 20, 445 (1998).

33. L. R. Squire, *Psychol. Rev.* 99, 195 (1992); D. L. Schacter and E. Tulving, Eds., *Memory Systems* (MIT Press, Cambridge, MA, 1994).

34. M. Mishkin, *Philos. Trans. R. Soc. London, Ser. B* 298, 85 (1982); L. R. Squire and S. Zola-Morgan, *Science* 253, 1380 (1991).

35. J. O'Keefe and J. Dostrovsky, *Brain Res.* 34, 171 (1971); H. Eichenbaum et al., *Neuron* 23, 209 (1999).

36. L. R. Squire and E. R. Kandel, *Memory: From Mind to Molecules* (Scientific American Library, New York, 1999); H. Eichenbaum, *Nature Rev. Neurosci.* 1, 1 (2000); P. Goldman-Rakic, *Philos. Trans. R. Soc. London, Ser. B* 351, 1445 (1996); R. Desimone, *Proc. Natl. Acad. Sci. U.S.A.* 93, 13494 (1996); S. Higuchi and Y. Miyashita, *Proc. Natl. Acad. Sci. U.S.A.* 93, 739 (1996).

37. R. F. Thompson and D. J. Krupa, *Annu. Rev. Neurosci.* 17, 519 (1994); J. LeDoux, *The Emotional Brain* (Simon & Schuster, New York, 1996); J. L. McGaugh, *Science* 287, 248 (2000); M. Mishkin et al., in *Neurobiology of Learning and Memory,* G. Lynch et al., Eds. (Guilford, New York, 1984), pp. 65–77; D. L. Schacter and R. L. Buckner, *Neuron* 20, 185 (1998).

38. V. Castellucci et al., *Science* 167, 1745 (1970); M. Brunelli et al., *Science* 194, 1178 (1976); C. Bailey et al., *Proc. Nat. Acad. Sci. U.S.A.* 93, 13445 (1996).

39. S. Benzer, *Sci. Am.* 229, 24 (1973); W. G. Quinn and R. J. Greenspan, *Annu. Rev. Neurosci.* 7, 67 (1984); R. L. Davis, *Physiol. Rev.* 76, 299 (1996); J. Yin and T. Tully, *Curr. Opin. Neurobiol.* 6, 264 (1996).

40. D. O. Hebb, *The Organization of Behavior: A Neuropsychological Theory* (Wiley, New York, 1949).

41. T. V. P. Bliss and T. Lomo, *J. Physiol. (Lond.)* 232, 331 (1973).

42. M. Mayford et al., *Science* 274, 1678 (1996); J. Tsien et al., *Cell* 87, 1327 (1996); A. Silva et al., *Annu. Rev. Neurosci.* 21, 127 (1998); S. Martin et al., *Annu. Rev. Neurosci.* 23, 613 (2000); E. P. Huang and C. F. Stevens, *Essays Biochem.* 33, 165 (1998); R. C. Malenka and R. A. Nicoll, *Science* 285, 1870 (1999); H. Korn and D. Faber, *CR Acad. Sci. III* 321, 125 (1998).

43. W. T. Greenough and C. H. Bailey, *Trends Neurosci.* 11, 142 (1998).

44. D. V. Buonomano and M. M. Merzenich, *Annu. Rev. Neurosci.* 21, 149 (1998); C. Gilbert, *Proc. Nat. Acad. Sci. U.S.A.* 93 10546 (1996); T. Elbert et al., *Science* 270, 305 (1995).

45. The work of E.R.K. is supported by NIMH, the G. Harold and Leila Y. Mathers Foundation, the Lieber Center for Research on Schizophrenia, and the Howard Hughes Medical Institute. The work of L.R.S. is supported by the Medical Research Service of the Department of Veterans Affairs, NIMH, and the Metropolitan Life Foundation. We thank Thomas Jessell and Thomas Albright for their helpful comments on the manuscript.

12

Piecing Together the Biggest Puzzle of All

MARTIN J. REES

In a fitting final essay, Martin J. Rees celebrates the ways astronomers and cosmologists have systematically uncovered the biography of the universe. Rife with neutron stars, black holes, and the possibility of multiple universes that emerge from quantum fluctuations, it's a story as grand and awesome as it is strange.

MARTIN J. REES is Royal Society Research Professor, a fellow of King's College in Cambridge, England, and England's Astronomer Royal at the Royal Observatory in Greenwich. He was formerly professor at Sussex University and director of the Institute of Astronomy at Cambridge. Besides his extensive research in astrophysics and cosmology, Rees regularly writes books and articles for a general readership.

Timeline of Astrophysics and Cosmology

1054
Chinese astronomers observe a "guest star," the remnants of which are now known as the Crab Nebula.

Ninth to eleventh centuries
Arabic and Persian astronomy flourishes, yielding star charts and catalogs, chronicles of planetary movements, and other fundamental observations.

1543
Nicolaus Copernicus champions a heliocentric world system to supercede the geocentric system.

1608
Hans Lippershey manufactures the first telescopes, which originally were used for military purposes. Two years later Galileo Galilei uses his own telescopes to discover moons of Jupiter, Saturn's "odd handles," the phases of Venus, and sunspots.

Throughout human history, our existence and our place in nature have been enduring mysteries. Only during the twentieth century have astronomers and cosmologists fully realized the scale of our cosmos and understood the physical laws that govern it. This story sets our Earth in an evolutionary context stretching back before the birth of our solar system—right back, indeed, to the primordial event that set our entire cosmos expanding about 12 billion years ago. Quasars, black holes, neutron stars, and the big bang have entered the general vocabulary, if not the common understanding. Fundamental questions about our universe that were formerly in the realm of speculation are now within the framework of empirical science.

Partial view of immensity. The eighteenth-century polymath Thomas Wright imagined that God had created multiple world systems. Thomas Wright, "An Original Theory of New Hypothesis of the Universe" (1750), A Facsimile Reprint *(MacDonald, London, and Elsevier, New York, 1971).*

Gravity, General Relativity, Neutron Stars, and Black Holes

Central to many of these questions is gravity. It's the governing force in the cosmos. It holds planets in their orbits, binds stars and galaxies, and determines the fate of our universe. Isaac Newton's seventeenth-century theoretical description of gravity remains accurate enough to program the trajectories of spacecraft on their journeys to Mars, Jupiter, and beyond. But ever since 1905, when Albert Einstein's special theory of relativity showed that instantaneous transfer of information was precluded, physicists accepted that Newton's laws would be inadequate when the motions induced by gravity approached the speed of light. Einstein's general relativity (published in 1916), however, copes quite consistently even with situations when gravity is overwhelmingly strong.

General relativity ranks as one of the two pillars of twentieth-century physics; the other is quantum theory, a conceptual revolution that presaged our modern understanding of atoms and their nuclei. Einstein's intellectual feat was especially astonishing because, unlike the pioneers of quantum theory, he wasn't motivated by any experimental enigma.

It took 50 years before astronomers discovered objects whose gravity was strong enough to manifest the most distinctive and dramatic features of Einstein's theory. In the early 1960s, ultraluminous objects (quasars) were detected. They seemed to require an even more efficient power supply than nuclear fusion, the process by which stars shine; gravitational collapse seemed the most attractive explanation. The American theorist Thomas Gold expressed the exhilaration of theorists at that time. In an after-dinner speech at the first big conference on the new subject of relativistic astrophysics, held in Dallas in 1963, he said:

> The relativists with their sophisticated work [are] not only magnificent cultural ornaments but might actually be useful to science! Everyone is pleased: the relativists, who feel they are being appreciated, who are suddenly experts in a field they hardly knew existed; the astrophysicists for having enlarged their domain. . . . It is all very pleasing, so let us hope it is right.

Observation—using the novel techniques of radio and X-ray astronomy—bore out Gold's optimism. In the 1950s, the world's best optical telescopes were concentrated in the United States, particularly in California. This shift from Europe had come about for climatic as well as financial reasons. However, radio waves from space can pass through clouds, so Europe (and

1668
Isaac Newton designs and builds the first reflecting telescope.

1687
Isaac Newton publishes his *Principia Mathematica,* in which he lays out mathematically the laws of motion, universal gravitation, and the scientific method in general.

1755
Immanuel Kant proposes that the solar system arose from a vast nebula of material. At the end of the century, Pierre-Simon de Laplace develops the theory further, with more mechanistic details about stellar evolution.

1781
In his search for comets, Charles Messier catalogs tens of "deep sky" objects, although no one yet can interpret them as galaxies, nebulae, and star clusters.

1789
William Herschel completes a telescope with a 49-inch mirror, with which he can resolve stars in different nebulae.

1802
William Wollaston discovers dark lines in solar spectrum; 12 years later Joseph von Fraunhofer realizes that such lines can be used for spectroscopic studies.

1838
Friedrich Bessel is the first to publish the distance of a star, 61 Cygni.

1842
Christian Doppler describes the wavelength-altering effect that becomes known by his name.

1859
Gustav Kirchhoff and Robert Bunsen introduce spectroscopy to chemistry and use it to identify elements present in the sun.

1904
The era of large-scale telescopes begins with the construction of the Mount Wilson Observatory.

Australia) were able to develop the new science of radio astronomy without any climatic handicap.

Some of the strongest sources of cosmic radio noise could be readily identified. One was the Crab Nebula, the expanding debris of a supernova explosion witnessed by Asian astronomers in A.D. 1054. Other sources were remote extragalactic objects, involving, we now realize, energy generation around gigantic black holes. These detections were unexpected. The physical processes that emit the radio waves, although now well understood, were not predicted.

The most remarkable serendipitous achievement of radio astronomy was the discovery of neutron stars in 1967 by Anthony Hewish and Jocelyn Bell. These stars are the dense remnants left behind at the core of some supernova explosions. They were detected as pulsars: They spin (sometimes many times per second) and emit an intense beam of radio waves that sweeps across our line of sight once per revolution. The importance of neutron stars lies in their extreme physics—colossal densities, strong magnetic fields, and intense gravity.

In 1969, a very fast (30-Hz) pulsar was found at the center of the Crab Nebula. Careful observations showed that the pulse rate was gradually slowing. This was natural if energy stored in the star's spin was being gradually converted into a wind of particles, which keeps the nebula shining in blue light. Interestingly, this pulsar's repetition rate, 30 per second, is so high that the eye sees it as a steady source. Had it been equally bright but spinning more slowly—say, 10 times a second—the remarkable properties of this little star could have been discovered 70 years ago. How would the course of twentieth-century physics have been changed if superdense matter had been detected in the 1920s, before neutrons were discovered on Earth? One cannot guess, except that astronomy's importance for fundamental physics would surely have been recognized far sooner.

Neutron stars were found by accident. No one expected them to emit strong and distinctive radio pulses. If theorists in the early 1960s had been asked the best way to detect a neutron star, most would have suggested a search for X-ray emission. After all, if neutron stars radiate as much energy as ordinary stars, but from a much smaller surface, they must be hot enough that the radiation from them is in X-rays. So it was X-ray astronomers who appeared best placed to discover neutron stars.

X rays from cosmic objects get absorbed by Earth's atmosphere, however, and so can only be observed from space. X-ray astronomy, like radio astronomy, received its impetus from wartime technologies and expertise. In this case it was scientists in the United States who took the lead, especially the late Herbert Friedman and his colleagues at the U.S. Naval Research Laboratory. Their first X-ray detectors, mounted on rockets, each yielded only a few minutes of useful data before they crashed back to the ground. X-ray astronomy spurted forward in 1970 when NASA launched the first X-ray satellite, which could gather data for years at a time. Through this project and its many successors, X-ray astronomy has proved itself to be a crucial new window on the universe.

X rays are emitted by unusually hot gas and by especially intense sources. X-ray maps of the sky consequently highlight the hottest and most energetic objects in the cosmos. Neutron stars, which pack at least as much mass as the sun in a volume little more than 10 kilometers in diameter, are among these. Their gravity is so strong that relativistic corrections amount to 30 percent.

Some stellar remnants, we now suspect, collapse beyond neutron-star densities to form black holes, which distort space and time even more than a neutron star does. An astronaut who ventured within the horizon around a black hole could not transmit any light signals to the external world—it is as though space itself is being sucked inward faster than light moves out through it. An external observer would never witness the astronaut's final fate: Any clock would appear to run slower and slower as it fell inward, so the astronaut would appear impaled at the horizon, frozen in time.

The Russian theorists Yakov Zeldovich and Igor Novikov, who studied how time was distorted near collapsed objects, coined the term "frozen stars" in the early 1960s. The term "black hole" was coined in 1968, when John Wheeler described how "light and particles incident from outside . . . go down the black hole only to add to its mass and increase its gravitational attraction."

The black holes that represent the final evolutionary state of stars have radii of 10 to 50 kilometers. But there is now compelling evidence that holes weighing as much as millions, or even billions, of suns exist at the centers of most galaxies. Some of them manifest themselves as quasars—concentrations of energy that outshine all the stars in their host galaxy—or as intense sources of cosmic radio emission. Others, including one at our own galactic center, are quiescent, but affect the orbits of stars passing close to them.

Viewed from outside, black holes are standardized objects: No traces persist to distinguish how a particular hole formed, nor what objects it swallowed. In 1963 the New Zealander Roy Kerr discovered a solution of Einstein's equations that represented a collapsed rotating object. The "Kerr solution" acquired paramount importance when theorists realized that it describes space-time around any black hole. A collapsing object quickly settles down to

1905
Albert Einstein lays down the basis of his special theory of relativity, in which he states that physical laws are the same in all inertial reference frames and that the speed of light in a vacuum is a universal constant.

1916
Einstein articulates his general theory of relativity, in which gravity is interpreted in the context of curved space-time.

1917
Einstein adds the "cosmological constant" to his equations.

1918
Harlow Shapley proposes his model of galaxy structure.

1920
Vesto Slipher reports his discovery that spectra of galaxies are redshifted.

1924
Edwin Hubble proves that galaxies reside beyond the Milky Way.

Aleksandr Friedmann proposes the first "big bang" solution of Einstein's equations.

1927
Georges Lemaître suggests that there was a creation event that led to an expanding universe governed by Einstein's field equations.

1929
Using redshift data and galaxy distances, Edwin Hubble lays groundwork for the idea that the universe is expanding.

1931
Karl Jansky discovers cosmic radio waves and initiates the era of radio astronomy.

1937
Grote Reber constructs the first radio telescope, which has a 9.4-meter diameter.

1948
Ralph Alpher, Hans Bethe, and George Gamow examine elemental synthesis in a rapidly expanding universe.

Fred Hoyle, Thomas Gold, and Hermann Bondi propose steady-state cosmology.

a standardized stationary state characterized by just two numbers: those that measure its mass and its spin. Roger Penrose, the mathematical physicist who perhaps did the most to stimulate the renaissance in relativity theory in the 1960s, has remarked, "It is ironic that the astrophysical object which is strangest and least familiar, the black hole, should be the one for which our theoretical picture is most complete."

The discovery of black holes opened the way to testing the most remarkable consequences of Einstein's theory. The radiation from such objects comes primarily from hot gas swirling downward into the "gravitational pit." It displays huge Doppler effects, as well as having an extra redshift because of the strong gravity. Spectroscopic study of this radiation, especially the X Rays, can probe the flow very close to the hole and diagnose whether the shape of space near it agrees with what theory predicts.

The Expanding Cosmos

Our own Milky Way contains about 100 billion stars, mainly in a disc orbiting a central hub. Nothing was known about any more remote parts of the universe until the 1920s, but it's now recognized that our galaxy is one among billions.

Galaxies occur mostly in groups or clusters, held together by gravity. Our own Local Group, a few million light-years across, contains the Milky Way and Andromeda, together with 34 smaller galaxies. This group is near the edge of the Virgo Cluster, an archipelago of several hundred galaxies, whose core lies about 50 million light-years away. Clusters and groups are themselves organized in still larger aggregates. The so-called Great Wall, a sheetlike array of galaxies about 200 million light-years away, is the nearest and most prominent of these giant features.

Perhaps the most important single broad-brush fact about our universe is that all galaxies (except for a few nearby galaxies in our own cluster) are receding from us. Moreover, the redshift—a measure of the recession speed—is larger for the fainter and more distant galaxies. We seem to be in an expanding universe where clusters of galaxies get more widely separated—more thinly separated through space—as time goes on.

The simple relation between redshift and distance is named after Edwin Hubble, who first claimed such a law in 1929. Hubble could only study relatively nearby galaxies, whose recession speeds were less than 1 percent of the speed of light. Thanks to technical advances and larger telescopes, the data

Hubble's vision. Edwin Hubble, shown here with the 38-inch Schmidt Telescope at the Palomar Observatory, discerned the connection between an object's redshift and its distance. *Palomar Observatory/American Institute of Physics/Emilio Segre Visual Archives.*

now extend to galaxies whose apparent recession amounts to a good fraction of the speed of light. It is conceptually preferable to attribute the redshift to "stretching" of space while the light travels through it. The amount of redshifting—in other words, the amount the wavelengths are stretched—tells us how much the universe has expanded while the light has been traveling toward us.

Models for an expanding, homogeneous universe, some based on Einstein's general relativity, had been devised in the 1920s and 1930s. But there was then little quantitative evidence on the extent to which our universe was actually homogeneous. Still less was it possible to discriminate between alternative models.

Astronomers are now mapping many more clusters like Virgo and more features like the Great Wall. But deeper surveys don't seem to reveal anything even larger. A box 200 million light-years on a side (a distance still small compared to the horizon of our observations, which is about 10 billion light-years away) can accommodate the largest aggregates. Such a box, wherever it was

1958

James Van Allen pioneers satellite-based studies when he uses data from a particle counter on *Explorer IV* to discover Earth's magnetosphere.

Martin Ryle provides the first strong evidence that there were more radio galaxies in the past, implying that the universe is not in a steady state.

1960s

The Great Epoch of planetary exploration begins; eventually every object in the solar system bigger than the moon is visited by satellites.

1963

Roy Kerr discovers a solution of Einstein's equations, which represents a collapsed rotating object.

Maarten Schmidt is the first to recognize a quasar.

1965

Arno Penzias and Robert Wilson discover cosmic background radiation, which provides strong evidence for the big bang theory.

1967

Anthony Hewish and Jocelyn Bell detect the first radio pulsar.

1970s

Satellite-based studies in the X-ray, ultraviolet, and other nonvisible regions of the spectrum get under way.

1974

The case for dark matter (which was first advanced by Fritz Zwicky in the 1930s) becomes widely accepted.

1980s

The concept of inflation, first clearly proposed by Alan Guth, suggests an explanation for the scale and expansion of the universe. An influx of particle physicists into cosmology stimulates much speculative research on the ultraearly universe.

The era of computer simulations of galaxy formation begins.

1990

Astronauts aboard the space shuttle *Discovery* place the Hubble Space Telescope into orbit. Three years later, another shuttle crew fits the Hubble with corrective optics.

placed in the universe, would contain roughly the same number of galaxies, grouped in a statistically similar way, into clusters, filamentary structures, and so on.

Even the very biggest conspicuous cosmic structures are small compared with the largest distances our telescopes can probe. This makes the science of cosmology possible, by allowing us to define the average properties of our universe and to use simple, homogeneous models as a valid approximation.

In the 1950s, Allan Sandage was a lone pioneer in advocating how the 200-inch (5-m) Mount Palomar telescope could probe deep enough into space and, therefore, far enough back in the past, to test cosmological models. To detect changes in the expansion rate, or evolution in the galaxy population, it is necessary to look at objects so far away that their light set out billions of years ago.

In the last 40 years, the development of ever more capable and revealing telescopes and observational techniques has made this possible. Over a dozen telescopes with four-meter mirrors were built during the 1970s and 1980s. Replacement of photographic plates, with 1 percent quantum efficiency, by solid-state detectors with efficiencies up to 80 percent hugely enhanced detectability of faint and distant objects. A new generation of still larger telescopes (of which the two Keck Telescopes in Hawaii are the first) is now coming into service. Perhaps most impressive of all is the unimaginatively named Very Large Telescope, a cluster of four telescopes, each with an 8.2-meter mirror, constructed in the Chilean Andes by a consortium of European nations. This instrument not only collects more light than any previous telescope but is also intended to yield sharper images, by compensating for the fluctuations in the atmosphere and linking the telescopes together so they can function as an interferometer.

There also have been dramatic advances in space-based observations. Although initially dogged by delays, flaws, and cost overruns, the Hubble Space Telescope has been fulfilling the hopes astronomers had for it. The Hubble Deep Field images—obtained by pointing steadily for several days at a small patch of sky—reveal literally hundreds of faint smudges, even within a field of view so small that it would cover less than 1/100 the area of the full moon. Each smudge is an entire galaxy, thousands of light-years in size, which appears so small and faint because of its huge distance. We are viewing these remote galaxies at a very primitive evolutionary stage. They have as yet no complex chemistry: There would have been very little oxygen, carbon, and other elements to make planets, and so scant chance of life.

We now have snapshots that take us billions of years back in time, to the era when the first galaxies were forming. The first stars may actually have formed even earlier, in aggregates smaller than present-day galaxies, which are too faint for even the largest existing telescopes to reveal.

1990–94
Data from NASA's Cosmic Background Explorer satellite show that the background spectrum obeys a "blackbody" or thermal law, with a precision of one part in 10,000, and also reveals nonuniformities in the temperature at a level of one part in 100,000.

"Fossils" of the Hot Beginning

What about still more remote epochs, before even the first stars had formed? In the later 1920s, the MIT–trained Belgian priest Georges Lemaître, along with Aleksandr Friedmann in Russia, was a pioneer of the idea that everything began in a dense state and that its structure unfolded as it expanded. Lemaître wrote: "The evolution of the universe can be likened to a display of fireworks that has just ended: some few wisps, ashes and smoke. Standing on a well-chilled cinder we see the fading of the suns, and try to recall the vanished brilliance of the origin of the worlds."

Big bang. In 1927, Georges Lemaitre suggested that there was a creation event that led to an expanding universe governed by Einstein's field equations. *American Institute of Physics/Emilio Segre Visual Archives.*

This "vanished brilliance" was revealed in 1965. Arno Penzias and Robert Wilson, two scientists at the Bell Telephone Laboratories striving to reduce the noise in an antenna in Holmdel, New Jersey, serendipitously discovered that all space is slightly warmed by microwaves with no apparent source. In 1990, John Mather and his colleagues, using NASA's Cosmic Background Explorer (COBE) satellite, showed that the spectrum obeys a "blackbody" or thermal law, with a precision of one part in 10,000—just what would be expected if it were indeed a relic of a "fireball" phase when everything in our universe was squeezed intensely hot, dense, and opaque. The cosmic expansion would have cooled and diluted the original radiation and stretched its wavelength, but it would still be around, pervading all of space.

1992
Alexander Wolszczan and Dale Frail discover the first planets beyond the solar system, orbiting a pulsar.

The present background temperature is only 2.728 degrees above absolute zero, but this represents a surprising amount of heat: 412 million quanta of radiation (photons) in each cubic meter of the present-day universe. In contrast, all the observed stars and gas in the universe, if spread uniformly

Big noise. In 1965, Arno Penzias (left) and Robert Wilson discovered cosmic background radiation, which provides strong evidence for the big bang theory. *American Institute of Physics/Emilio Segre Visual Archive.*

1994
The first Keck telescope inaugurates a new era in the study of the high-redshift universe.

1995
Michel Mayor and Didier Queloz discover the first planet around a sunlike star, 51 Pegasi. An era of extrasolar planetary discovery begins in earnest.

1998–2000
Evidence seems to converge on a specific set of parameters to describe our universe.

through space, would amount to only about 0.2 atoms per cubic meter—more than a billion times less than the photon density.

According to the big bang theory, everything would once have been compressed hotter than the center of a star—certainly hot enough for nuclear reactions. The most important of these reactions happen at a temperature of roughly a billion degrees. However, the universe cooled below this temperature within three minutes, and (fortunately for us) this didn't allow time to process primordial material into iron, as in the hottest stars—nor even into carbon, oxygen, etc.

This was contrary to George Gamow's conjecture that the entire Periodic Table was "cooked" in the early universe. In the 1950s, Fred Hoyle, William Fowler, and Geoffrey and Margaret Burbidge—and, in parallel work, Alistair Cameron—developed an alternative scheme, which quantitatively explained almost the entire Periodic Table as the outcome of nuclear fusion in stars and supernovae. After later refinements, the calculated mix of atoms is gratifyingly close to the proportions now observed.

The oldest stars, which would have formed from gas early in galactic history, when it was less "polluted," are indeed deficient in heavy elements, just as stellar nucleosynthesis theory would lead one to expect. However, even the oldest objects turned out to be 23 to 24 percent helium: No star, galaxy, or nebula has been found where helium is less abundant than this. It seems as though the galaxy started not as pure hydrogen but was already a mix of hydrogen and helium.

The "hot big bang" theory neatly solves this mystery. Reactions in the hot early phases would turn about 23 percent of the hydrogen into helium, but the universe cooled down so fast that there wasn't time to synthesize the elements higher up the Periodic Table (apart from a trace of lithium). Attributing most cosmic helium to the big bang thus solved a long-standing problem—why there is so much of it, and why it is so uniform in its abundance—and emboldened cosmologists to take the first few seconds of cosmic history seriously.

Another product of the big bang is deuterium (heavy hydrogen). Deuterium's abundance relative to hydrogen was until recently uncertain. But the ratio, measured in Jupiter, in interstellar gas, and in remote intergalactic clouds, is now pinned down to be about 1/50,000. The origin of even this trace poses a problem, because deuterium is destroyed rather than created in stars. As a nuclear fuel, it is easier to ignite than ordinary hydrogen, so newly formed stars destroy their deuterium during their initial contraction, before settling down in their long, hydrogen-burning phases.

If we assume a present universe-wide average density of 0.2 atoms per cubic meter and compute what mixture of atoms would emerge from the cooling fireball, we find that the proportions of hydrogen, deuterium, and helium (and, as a bonus, lithium as well) agree with observations. This is gratifying, because the observed abundances could have been entirely out of line with the predictions of any big bang; or they might have been consistent, but only for a density that was far below, or else far above, the range allowed by observation.

The Emergence of Structure

If our universe started off as a hot, amorphous fireball, how did it differentiate into the observed pattern of stars, galaxies, and clusters? This is actually a natural outcome of the workings of gravity, which leverages, over time, even very slight initial irregularities into conspicuous density contrasts.

Theorists can now follow virtual universes in a computer. Slight fluctuations are fed in at the start of the simulation. As the universe expands, incipient galaxies and larger structures emerge and evolve. The purely gravitational aspects of this process can be modeled fairly well. However, when gas contracts under gravity into a protogalaxy, its density has to rise by many powers of 10 before stars form, and complicated dynamics and radiative transfer determine what their masses are. Moreover, the energy output from the first stars exerts uncertain feedback on what happens later. This is too complicated to be computed and has to be approximated by plausible "recipes," chosen in the light of local observations.

Despite these limitations, simulations of structure formation have achieved remarkable success in accounting for present-day morphology of galaxies and clusters. Moreover, these models can be checked by seeing how well they account also for the new high-redshift data, which tell us what the universe was like at earlier times.

There is another consistency check on computed scenarios. In these simulations, the predicted sizes and clustering of galaxies depend on the form and amplitude of the initial fluctuations. The microwave background should bear the imprint of these fluctuations and thus offers an independent line of evidence on their amplitude. This radiation in effect comes from a surface so far away that it is being observed at an era when the fluctuations were still of small amplitude. Radiation from an incipient cluster on that surface would appear slightly cooler, because it loses extra energy climbing out of the gravitational pull of an overdense region; conversely, radiation from the direction of an incipient void would be slightly hotter. The predicted fractional differences in the temperature over the sky due to this effect are about one part in 10^5. On some scales, a somewhat larger Doppler fluctuation is predicted, due to the associated motions.

Measuring one part in 100,000 of background radiation, which is itself 100 times cooler than the atmosphere, is a daunting technical challenge. But it has now been achieved. Fluctuations were first measured by George Smoot and his colleagues, using four years of data gathered by the COBE satellite. These measurements were restricted to angular scales exceeding seven degrees, however. They since have been supplemented and extended by ground-based and balloon experiments. The amplitudes are indeed consistent with what is required for galaxy formation. Within a few years, two new spacecraft—NASA's Microwave Anisotropy Probe and the European Space Agency's Planck satellite—will yield data precise enough to settle many key questions about cosmology, the early universe, and how galaxies emerged.

Dark Matter, Omega, and the First Microsecond

In about five billion years, the sun will die and Earth with it. At about the same time (give or take a billion years) the Andromeda galaxy, already falling toward us, may crash into our own Milky Way. But will the universe go on expanding forever? Or will the entire firmament eventually recollapse to a big crunch, where everything will suffer the same fate as an unwary astronaut who falls into a black hole?

The answer depends on how much the cosmic expansion is being decelerated by the gravitational pull that everything exerts on everything else. It is straightforward to calculate that the expansion can eventually be reversed if there is, on average, more than about five baryons in each cubic meter—the so-called critical density. (A baryon is the collective term for protons and neutrons, the heavy ingredients of all atoms.) That doesn't sound like much. But if all the galaxies were dismantled, and their constituent stars and gas spread uniformly through space, they'd make an even emptier vacuum—one baryon in every 10 cubic meters. Add to that what seems to be a similar amount of material in diffuse intergalactic gas, and the resulting density amounts to 0.2 baryons per cubic meter.

That's 25 times less than the critical density, which at first sight implies perpetual expansion. But the actual situation is less straightforward. Astronomers have discovered that galaxies, and even entire galaxy clusters, would fly apart unless held together by the gravitational pull of five to 10 times more material than we actually see. This is the famous "dark matter" mystery.

There are many candidates for dark matter. Earlier ideas included very faint stars (known as "brown dwarfs"), or the remnants of massive stars. However, most cosmologists suspect that dark matter is mainly exotic particles left over from the big bang and is not made of baryons at all.

There are two main reasons for this belief. First, the proportions of helium and deuterium that are calculated to emerge from the big bang are sensitive to the baryon density and would be discrepant with observations if the average baryonic density were, say, between 1 and 2, rather than 0.2 per cubic meter. Extra dark matter in exotic particles that do not participate in nuclear reactions, however, would not scupper the concordance with a baryon density of only 0.2 per cubic meter.

Second, the formation of galaxies would be hard to understand if the bulk of their mass were baryonic. Nonbaryonic matter can cluster more efficiently in the early universe because it does not feel the opposing influence of radiation pressure. If the universe were purely baryonic, it would be hard to reconcile its present highly structured state with the small amplitude of the primordial fluctuations implied by the microwave background anisotropies.

The dark matter mystery may yield to a three-pronged attack:

1. *Direct detection.* Using sensitive detectors in underground laboratories, searches are now under way for dark matter candidates, including heavy neutral particles and axions.

2. *Progress in particle physics.* If we knew more about the types of particles that could exist in the ultraearly universe, then we could confidently calculate how many should survive from the first microseconds of the big bang and how much dark mass they would contribute.

3. *Simulations of galaxy formation and large-scale structure.* When and how galaxies form and cluster depends on what their gravitationally dominant constituent is. It is possible to simulate the formation of galaxies on a computer, making alternative assumptions about the dark matter. If one assumption yielded an outcome that matched real galaxies especially closely, this would at least be corroborative evidence for one candidate rather than another.

Cosmologists denote the ratio of the actual density to the critical density by Ω. There is certainly enough dark matter to make $\Omega = 0.2$. Until recently, we couldn't rule out several times this amount—comprising the full critical density, $\Omega = 1$—in the space between clusters of galaxies. But it now seems that, in toto, atoms and dark matter don't contribute more than about 30 percent of critical density.

We can never be sure of the long-term future: New physics may intervene, and domains beyond our observational range may be different from the part of the universe we can see. But with those provisos, the odds favor perpetual expansion. The galaxies will disperse, and they will fade as their stars all die and their material gets locked up in old white dwarfs, neutron stars, and black holes.

There is, moreover, tantalizing evidence for an extra repulsion force that overwhelms gravity on the cosmic scale. Studies of the redshift versus the apparent brightness of distant supernovae suggest that galaxies may disperse at an accelerating rate. *Science* rated this—perhaps prematurely—as the most important discovery of 1998, in any field. If this work is corroborated, the forecast is an even emptier universe. All galaxies beyond our local group will accelerate away, disappearing completely from view as their redshift rises exponentially toward infinity.

The idea of a cosmological repulsion goes back to 1917, when Einstein introduced an extra term into his equations, which he called the cosmological constant, or 1. His motive was to allow a static universe, in which the repulsive force counteracted gravity. He later abandoned the idea, calling it his "biggest blunder," when Hubble discovered that the universe was actually expanding. However, from our modern perspective, 1 can be envisaged as dark energy latent in empty space. It leads to a repulsion because, according to Einstein's equation, gravity depends on pressure as well as density, and if the pressure is sufficiently negative (as it has to be for vacuum energy), the net effect is repulsive.*

The cosmological constant corresponds to a vacuum energy that is unchanging as the universe expands. Cosmologists have recently suggested variants—forms of dark energy, dubbed quintessence, with negative pressure that could gradually decay.

There is another line of evidence for some form of dark energy, quite independent of the indications from supernovae for an accelerating universe. Theory tells us that fluctuations or ripples in the microwave background should be biggest on a particular length scale that is related to the maximum distance a sound wave can travel. The angular scale corresponding to this length depends, however, on the geometry of the universe.

Measurements have now pinned down the angular scale of this Doppler peak with better than 10 percent precision. The results are consistent with a "flat" universe. In contrast, if there were no mass-energy apart from enough baryons and dark matter to yield 0.2 to 0.3 for Ω, we would be in an open universe, where this angle would be smaller by a factor of 2—definitely in conflict with observations. Reconciliation with the microwave background measurements would be achieved if dark energy made up the balance, so that the universe was, after all, flat; the dominant dark energy would then drive the accelerating expansion.

Within just the last two years, a remarkable concordance among several seemingly independent observations has emerged, leading to a preferred choice for the key parameters describing our universe. It seems that the universe is flat, with baryons providing 4 percent of the mass-energy, dark mat-

---

*The gravitational attraction depends not on density alone but on (density) + 3 (pressure)/c^2. The pressure associated with vacuum energy must be negative, as the following simple argument explains. If a box of gas expands by an amount dV, then the energy inside is reduced by pdV, the work done by the pressure during the expansion. But if a box of vacuum contains energy, its energy in the box has gone up. The pdV term must therefore be negative, implying a pressure of $-c^2$ times the vacuum density.

ter 20 to 30 percent, and dark energy the rest, i.e., 66 to 76 percent. The flatness vindicates a natural prediction of the "inflationary theory" (discussed below). The expansion accelerates because dark energy (with negative pressure) is the dominant constituent. But there seems nothing natural about the actual split between the three different contributions.

Is there any explanation for these numbers? For that matter, why do simple cosmological models based on precise homogeneity and isotropy fit so well? A completely chaotic and irregular universe would at first sight seem more probable. If there is an answer to these questions, it lies in the initial instants of cosmic history.

For the last 20 years, cosmologists have suspected that the uniformity is a legacy of something remarkable that happened in the ultraearly universe: A very fierce cosmical repulsion, it is claimed, could have accelerated the expansion, so that a tiny patch of space-time expanded exponentially and homogenized when it was, perhaps, no more than 10^{-35} seconds old.

The generic idea that our universe inflated from something microscopic is compellingly attractive. Rather than assuming the expansion as an initial condition, it accounts for it physically. It looks like "something for nothing," but it isn't really. That's because our present vast universe may, in a sense, have zero net energy. Every atom has an energy because of its mass (Einstein's mc^2). But it has a negative energy due to gravity. We, for instance, are in a state of lower energy on Earth's surface than if we were up in space. And if we added up the negative potential energy we possess due to the gravitational field of everything else, it could cancel out our rest mass energy. Thus it doesn't, as it were, cost anything to expand the mass and energy in our universe.

Alan Guth spelled out this concept of inflation in 1981, building on work by others. It invokes extreme, untested physics and so still has unsure foundations. But it isn't just metaphysics. Its one generic prediction—that the universe would be stretched flat—seems to be borne out by observation. Moreover, observations can in principle firm it up. For instance, the slight ripples that are the seeds of galaxies and clusters could have been quantum fluctuations, imprinted when the entire universe was of microscopic size and stretched by inflationary expansion. The nonuniformities in the present universe depend, in a calculable way, on the physics of inflation, so observations will be able to probe this extreme physics and perhaps help us understand what caused inflation.

Some variants of the inflationary universe—espoused by Andrei Linde, Alex Vilenkin, and others—suggest that our big bang wasn't the only one. This fantastic speculation dramatically enlarges our concept of reality. The entire history of our universe might just be an episode, one facet, of the infinite multiverse. Current ideas on superstring theory, and the possibility of extra spatial dimensions, suggest other equally fascinating scenarios.

The Universe as We Are Coming to Know It

Cosmologists are no longer starved of data: Current progress is owed far more to observers and experimentalists than to armchair theorists. Actual physical probes remain confined to our own solar system, but vicarious exploration with telescopes and other techniques has extended our horizons. These tools allow us to study galaxies whose light has been journeying toward us for 90 percent of the time since the big bang.

And we are making sense of what we see. Stars follow finite life cycles whose main features are now quite well understood. There is no equivalent understanding of galaxies. But observations of nearby galaxies, and ones so remote that they are being viewed as they form, are helping.

We're witnessing a crescendo of discoveries that promises to continue throughout the next decade. It is something of a coincidence—of technology and funding, as well as of the way the intellectual discourse has developed—that there has, more or less concurrently, been an impetus on several fronts:

- Sensitive tools and techniques make it possible to intensively study the microwave background fluctuations.
- The Hubble Space Telescope (HST) has been fulfilling its potential for observing deep-space phenomena; new 8- to 10-meter telescopes are on line; new X-ray telescopes in space, and radio arrays on the ground, now offer greater sensitivity. And a decade from now, new space telescopes should carry the enterprise beyond what the HST can achieve.
- Large-scale clustering and dynamics studies, and big surveys of galaxies, will permit sensitive statistical tests that should discriminate among theories for structure formation.
- Dramatic advances in computer technology have allowed increasingly elaborate numerical simulations, which now incorporate realistic gas dynamics as well as gravity.
- New fundamental physics offers the hope of putting the ultraearly universe on as firm a footing as the later eras.

Should We Believe the Big Bang Scenario?

The extrapolation by astrophysicists and cosmologists back to a stage when the universe had been expanding for a few seconds deserves to be taken as seriously as, for instance, what geologists or paleontologists tell us about the early history of our Earth: Their inferences are just as indirect and generally less quantitative.

Moreover, there are several discoveries that might have been made over the last 30 years which would have invalidated the big bang hypothesis and which have not been made—the big bang theory has lived dangerously for decades and survived. Here are some of those absent observations:

- Astronomers might have found an object whose helium abundance was far below the amount predicted from the big bang—23 percent. This would have been fatal, because extra helium made in stars can readily boost helium above its pregalactic abundance, but there seems no way of converting all the helium back to hydrogen.
- The background radiation measured so accurately by the Cosmic Background Explorer satellite might have turned out to have a spectrum that differed from the expected "blackbody" or thermal form. What's more, the radiation temperature could have been so smooth over the whole sky that it was incompatible with the fluctuations needed to give rise to present-day structures like clusters of galaxies.
- A stable neutrino might have been discovered to have a mass in the range of 100 to 106 electron volts. This would have been fatal, because the hot early universe would have contained almost as many neutrinos as photons. If each neutrino weighed even a millionth as much as an atom, they would, in

toto, contribute too much mass to the present universe—more, even, than could be hidden in dark matter. Experimental physicists have been trying hard to measure neutrino masses, but they are seemingly too small to be important contributors to dark matter.

- The deuterium abundance could have been so high that it was inconsistent with big bang nucleosynthesis (or implied an unacceptably low baryon density).

The big bang theory's survival gives us confidence in extrapolating right back to the first few seconds of cosmic history and assuming that the laws of microphysics were the same then as now.

Some debates have been settled; some earlier issues are no longer controversial. As the consensus advances, new questions, which couldn't even have been posed earlier, are now being debated. Among them are: Why does our universe consist of the particular mix of baryonic matter and dark matter? What caused the initial favoritism for matter over antimatter? Is the dark matter composed of neutral particles surviving from the big bang, or is it something even more exotic? Are they the outcome of quantum fluctuations imprinted when our entire universe was of microscopic size? What is the mysterious dark energy that makes our universe flat, and how does this relate to inflation?

It will remain a challenge to understand the very beginning. This will require a new theory, perhaps a variant of superstrings, which unifies cosmos and microworld. Optimists hope for breakthroughs soon. But the aim of cosmology and astronomy is to map out how a simple fireball evolved, over 12 billion years, into the complex cosmic habitat we find around us. Understanding how the consequences of the basic laws have unfolded over cosmic history is an inexhaustible challenge for the new millennium.

References and Notes

N. A. Bahcall, J. P. Ostriker, S. Perlmutter, P. J. Steinhardt, *Science* 284, 1481 (1999).

M. C. Begelman and M. J. Rees, *Gravity's Fatal Attraction: Black Holes in the Universe* (W. H. Freeman & Company, New York, 1998).

R. Brawer and A. P. Lightman, *Origins: The Lives and Worlds of Modern Cosmologists* (Harvard University Press, Cambridge, MA, 1992).

E. R. Harrison, *Cosmology: The Science of the Universe* (Cambridge University Press, Cambridge, ed. 2, 2000).

J. A. Peakcock, *Cosmological Physics* (Cambridge University Press, Cambridge, 1998).

Reviews of Modern Physics 71(2), (1999), American Physical Society. Special centenary issue with many authoritative reviews of history and current status of subject.

INDEX